Anne Bayer

Spatial & temporal dynamics in a tar oil contaminated aquifer

Anne Bayer

Spatial & temporal dynamics in a tar oil contaminated aquifer

Focus on microbial activities

Südwestdeutscher Verlag für Hochschulschriften

Impressum/Imprint (nur für Deutschland/only for Germany)
Bibliografische Information der Deutschen Nationalbibliothek: Die Deutsche Nationalbibliothek verzeichnet diese Publikation in der Deutschen Nationalbibliografie; detaillierte bibliografische Daten sind im Internet über http://dnb.d-nb.de abrufbar.

Alle in diesem Buch genannten Marken und Produktnamen unterliegen warenzeichen-, marken- oder patentrechtlichem Schutz bzw. sind Warenzeichen oder eingetragene Warenzeichen der jeweiligen Inhaber. Die Wiedergabe von Marken, Produktnamen, Gebrauchsnamen, Handelsnamen, Warenbezeichnungen u.s.w. in diesem Werk berechtigt auch ohne besondere Kennzeichnung nicht zu der Annahme, dass solche Namen im Sinne der Warenzeichen- und Markenschutzgesetzgebung als frei zu betrachten wären und daher von jedermann benutzt werden dürften.

Coverbild: www.ingimage.com

Verlag: Südwestdeutscher Verlag für Hochschulschriften GmbH & Co. KG
Heinrich-Böcking-Str. 6-8, 66121 Saarbrücken, Deutschland
Telefon +49 681 37 20 271-1, Telefax +49 681 37 20 271-0
Email: info@svh-verlag.de

Approved by: München, TU, Diss., 2011

Herstellung in Deutschland (siehe letzte Seite)
ISBN: 978-3-8381-2353-0

Imprint (only for USA, GB)
Bibliographic information published by the Deutsche Nationalbibliothek: The Deutsche Nationalbibliothek lists this publication in the Deutsche Nationalbibliografie; detailed bibliographic data are available in the Internet at http://dnb.d-nb.de.

Any brand names and product names mentioned in this book are subject to trademark, brand or patent protection and are trademarks or registered trademarks of their respective holders. The use of brand names, product names, common names, trade names, product descriptions etc. even without a particular marking in this works is in no way to be construed to mean that such names may be regarded as unrestricted in respect of trademark and brand protection legislation and could thus be used by anyone.

Cover image: www.ingimage.com

Publisher: Südwestdeutscher Verlag für Hochschulschriften GmbH & Co. KG
Heinrich-Böcking-Str. 6-8, 66121 Saarbrücken, Germany
Phone +49 681 37 20 271-1, Fax +49 681 37 20 271-0
Email: info@svh-verlag.de

Printed in the U.S.A.
Printed in the U.K. by (see last page)
ISBN: 978-3-8381-2353-0

Copyright © 2012 by the author and Südwestdeutscher Verlag für Hochschulschriften GmbH & Co. KG and licensors
All rights reserved. Saarbrücken 2012

Spatial and temporal dynamics of microbial activities in a tar oil contaminated aquifer

Spucke nicht in den Brunnen, du wirst selbst daraus trinken müssen
(Russisches Sprichwort)

Zusammenfassung

Grundwasser dient als wichtige Trinkwasserquelle, ist allerdings sehr anfällig für anthropogene Verunreinigung. Eine Einleitung von organischen Schadstoffen, wie aromatische Kohlenwasserstoffe, in poröse Grundwasserleiter führt zur Entstehung von Schadstofffahnen. Ziel vorliegender Arbeit war die Untersuchung der zeitlichen und räumlichen Dynamik von physikalisch-chemischen Gradienten und mirkobiellen Abbauprozessen an einem ehemaligen Gaswerk-Standort in Düsseldorf-Flingern, der hauptsächlich mit BTEX (Benzol, Toluol, Ethylbenzol, Xylole) und Naphthalin verunreinigt ist. Desweiteren wurden Zonen mit aktivem Schadstoffaubbau lokalisiert, potenzielle abbaulimitierende Faktoren und die Auswirkungen der hydraulischen Schwankungen im Grundwasser auf den natürlichen Schadstoffabbau ermittelt. Ein weiterer Punkt war der Vergleich von Daten, die aus dem Grundwasser gewonnen wurden mit Daten aus dem Sediment. Im letzten Abschnitt der Arbeit wurde eine Methode für eine relativ schnelle direkte Quantifizierung der Zellzahl von Sedimentprokaryoten erfolgreich entwickelt.

Sowohl von Grundwasser- als auch von Sedimentproben wurden mehrere Parameter in einer vertikalen hohen Auflösung analysiert, um die vorherrschenden Redoxprozesse zu lokalisieren und zu charakterisieren. Ein Vergleich der wichtigsten geochemischen Daten des Grundwassers und des Sediments von 2006 zeigte eine ziemlich gute Übereinstimmung zwischen beiden Probearten. Dennoch wurde nur in den Wasserproben eine BTEX-Fahne direkt unterhalb des Grundwasserspiegels vorgefunden, während ein Großteil der polyzyklische aromatische Kohlenwasserstoffe (PAK) im Sediment in tiefer liegenden Zonen zu finden sind. Hinweise für Sulfatreduktion, worüber BTEX abgebaut wird, war in den Wasserproben sichtbar, wohingegen Redoxprozesse mit unlöslichen Elektronenakzeptoren, wie Eisenreduktion, hauptsächlich im Sediment gefunden wurden. Die Bodenproben haben eine größere Bedeutung als „Langzeit-Gedächtnis" während aus dem Grundwasser eher „Momentaufnahmen" gewonnen werden können.

Ein Ziel der durchgeführten Probennahmen war, die zeitlichen und räumlichen Dynamiken von physikalisch-chemischen Gradienten und mikrobiellen Abbauprozessen in Grundwassersystemen, die allgemein als relativ stabil gelten, aufzudecken. Es ist wichtig, die Auswirkungen der schwankenden hydraulischen Verhältnisse auf Diversität, Häufigkeit und Aktivität der mikrobiellen Gemeinschaft, zu kennen. Ein weiteres Ziel der vorliegenden Arbeit war das Wissen der aufeinander einwirkenden abiotischen und biotischen Faktoren zu erweitern, die Einfluss auf die Kapazität des biologischen Abbaus haben. Wir möchten wissen wer was, wo und unter welchen Bedingungen abbaut.

Beim Vergleich der Datensätze aller Probenahmen zwischen 2005 und 2009 konnten eindeutige Dynamiken der vertikalen Ausdehnung der Schadstofffahne beobachtet werden, auch wenn nur geringe Schwankungen des Grundwasserspiegels stattfanden. Einerseits wird das Potenzial der Selbstreinigungsprozesse (auch als natürliche Attenuation bezeichnet, NA) durch eine erhöhte Mischung gesteigert, andererseits kann die NA auch vermindert werden, wenn ungünstige Bedingungen zu den etablierten sessilen Abbauern gebracht werden.

2008 erreichte die Toluolkonzentration maximale Werte seit der Installation der hochauflösenden Multi-Level-Grundwassermessstelle (HR-MLW), wohingegen 2009 die Konzentrationen unerwartet gering ausfielen. Die hohe Toluolkonzentration kann einerseits durch eine mögliche zusätzliche Quelle erklärt werden, die aufgrund eines geringen Anstiegs des Wasserspiegels ausgewaschen wurde, oder durch einen Zusammenbruch des Toluol-Abbaus. Die stoffabhängige Analyse stabiler Isotope (CSIA) und auch Analysen der mikrobiellen Gemeinschaft deuten auf einen Kollaps des Toluolabbaus in 2008 mit Regeneration in 2009 hin. Außerdem konnten zum ersten Mal Schwankungen in der vertikalen BTEX-Verteilung mit dazu gehörendem Auftreten von Sulfid, einem Abbau-Produkt, nach nur zwei Wochen gemessen werden. Dies zeigt, dass es den Mikroorganismen möglich ist ihre Abbauaktivität relativ schnell zu adaptieren.

Mit der vorliegenden Arbeit konnte unser Wissen über die Komplexität der Prozesse erweitert werden, die den biologischen Abbau organischer Schadstoffe beeinflussen. Insbesondere hydraulische Eigenschaften poröser Grundwasserleiter führen zu deutlichen und unerwarteten Dynamiken des NA-Potentials im als Allgemein als stabil betrachteten Grundwassersystem. Es ist wichtig mehr über den zeitlichen Ablauf wesentlicher Störungen der NA zu wissen. Dies ermöglicht eine effizientere Umsetzung von Sanierungskonzepten, präzisere Vorhersagen über die Schadstoffverteilung und die Zeit, die ein Grundwasser-Ökosystem eines kontaminierten Standortes benötigt, um sich wieder zu erholen.

Das letzte Kapitel der vorliegenden Arbeit beschäftigt sich mit den Schwierigkeiten der direkten Quantifizierung von sediment-assoziierten Zellen. Dies ist wichtig, da der Großteil der Bakterien sich an der Oberfläche der Sedimentpartikel befindet. Um diese zu quantifizieren, wurde ein relativ schnelles und zuverlässiges Verfahren entwickelt, das am Sediment vom Standort Düsseldorf erfolgreich angewendet wurde. Des Weiteren wurde der Einfluss untersucht, den die Lagerung der Proben auf die Zellzahl, ATP-Menge und Anzahl der 16S rDNA-Gene hat. Hierbei konnte gezeigt werden, dass Einfrieren zu einer signifikanten Reduktion aller drei Parameter führt.

Abstract

Groundwater is a relevant source for drinking water, but is very vulnerable to anthropogenic pollution. Introduction of organic contaminants such as aromatic hydrocarbons into a porous aquifer leads to the formation of a contaminant plume. Aim of present work was the investigation of temporal and spatial dynamics of physical-chemical gradients and biodegradation processes at a former gasworks site in Düsseldorf-Flingern, contaminated mainly with BTEX (benzene, toluene, ethylbenzene, xylenes) and naphthalene. Furthermore, zones with active contaminant degradation were localized, potential limiting factors and the influence of hydraulic dynamics in groundwater on natural attenuation (NA) was identified. One further aspect was the comparison of data gained from groundwater samples with sediment samples. In the last part a reliable and relatively fast method for quantification of sediment prokaryotes was evaluated.

Groundwater as well as sediment samples multiple parameters were conducted in closely spaced vertical resolution to localize and characterize prevailing redox processes. Comparison of basic geochemical data of groundwater and sediment taken in 2006 showed a fairly good agreement between both sample types. Nevertheless, a BTEX plume could only be detected in groundwater samples directly beneath the groundwater table, whereas the bulk of polycyclic aromatic hydrocarbons (PAHs) was found at the sediment in deeper zones. Sulfate reduction involved in BTEX degradation was evident from groundwater samples, while redox processes including insoluble electron acceptors, such as iron reduction, were mainly detected in sediment. Sediment samples play a more important role as "long-term memory" whereas sampling of groundwater grants rather "snapshot" insights.

One aspect of the sampling campaigns was revealing the temporal and spatial dynamics of physical-chemical gradients and microbial processes in groundwater systems which are assumed to be relatively stable. It is important to know which effects transient hydraulic conditions have on the diversity, abundance and activity of the microbial community. A further aim of this study was the gain of additional knowledge concerning the interacting abiotic and biotic parameters which influence biodegradation capacity. We want to know who degrades what, where and under which conditions.

By comparing all data sets between 2005 and 2009, pronounced dynamics in the vertical plume expansion could be observed, even though only low fluctuations of the groundwater table took place. On the one hand, transient hydraulic conditions enhance the NA potential by increased mixing. On the other hand, NA can be decreased by groundwater fluctuations bringing unfavourable conditions to the established sessile degrader community.

In 2008, toluene concentrations reached a maximum since the installation of the high-resolution multi-level well (HR-MLW), whereas in 2009 the contaminant was unexpected low. The high toluene concentration can be explained by an additional source that was washed out by a slight elevation of the groundwater table, or by a collapse in toluene degradation. Compound specific isotope analysis (CSIA) as well as bacterial community analysis denote a collapse of toluene degradation in 2008 with regeneration in 2009. For the first time, short-term dynamics in the vertical BTEX distribution with inherent sulfide as degradation parameter were detectable after only two weeks of sampling interval. This shows that microorganisms have the possibility to adapt their degradation activity relatively fast.

With the present study our knowledge of the complexity of processes which are involved in biodegradation of organic contaminants could be increased. Particularly, hydraulic conditions within porous aquifers induce pronounced and unexpected dynamics to the NA potential in the regarded as stable groundwater system. It is important to enhance our knowledge of the timescale of intrinsic disturbances on NA. This enables a more efficient realization of remediation strategies, more precise predictions on contaminant distribution and on the time scale of aquifer ecosystem recovery at contaminated sites.

The last chapter of this work deals with the difficulties of enumeration of sediment associated cells. This is important, because the bulk of bacteria is associated to sediment particles. Hence, for quantification of these bacterial cells a relative fast and reliable method was developed and successfully applied to sediment samples of the Düsseldorf-Flingern aquifer. Moreover, the influence of sample storage on cell number, amount of ATP and number of 16S rDNA genes was investigated and showed significant reduction of all three parameters upon freezing.

x

Table of contents

Zusammenfassung v
Abstract viii
Table of contents xi
Abbreviations xiii
1. Introduction 1
   1.1. Groundwater – a vulnerable resource 1
   1.2. Toxicity 2
   1.3. Transport and distribution 3
   1.4. Natural attenuation 6
   1.5. Contaminant reduction – microbes as key players 7
   1.6. Evidence of biodegradation 9
   1.7. High resolution sampling at a gasworks site 11
   1.8. Dynamics in the environment influencing bacterial communities and biodegradation 13
   1.9. Motivation and outline of the thesis 14
   1.10. References 16
2. High resolution analysis of contaminated aquifer sediments and groundwater – what can be learned in terms of natural attenuation? 25
   2.1. Introduction 26
   2.2. Materials and methods 27
      2.2.1. Site description 27
      2.2.2. Groundwater sampling 28
      2.2.3. Sediment sampling 28
      2.2.4. Sample preparation and geochemical analysis 29
      2.2.5. Stable isotope analysis of sulfate 30
      2.2.6. Determination of bacterial cell numbers 31
      2.2.7. Bacterial community fingerprinting 31
   2.3. Results 32
      2.3.1. Distribution of contaminants 32
      2.3.2. Physical-chemical conditions and redox-specific parameters ... 34
      2.3.3. Direct cell counts and enzyme activities 35
      2.3.4. Depth-resolved analysis bacterial communities 37
   2.4. Discussion 40
      2.4.1. Small scale physical-chemical heterogeneities in groundwater and sediments 41
      2.4.2. Biodegradation via sulfate reduction 43
      2.4.3. Iron reduction and iron cycling processes 44
      2.4.4. Microbial patterns in groundwater and sediments 45
      2.4.5. Bacterial community shifts in sediment and water 47
   2.5. References 50

3. **Collapse and recovery of intrinsic toluene degradation – transient hydraulic conditions control natural attenuation in a tar oil contaminated sandy aquifer 55**

   3.1. Introduction 56
   3.2. Materials and methods 58
      3.2.1. Study site and groundwater sampling 58
      3.2.2. Physical-chemical characteristics of groundwater 58
      3.2.3. Quantification of cell numbers and bacterial carbon production (BCP) 58
      3.2.4. Stable isotope analysis 59
      3.2.5. Dynamic expansion of the BTEX plume 59
   3.3. Results and discussion 59
      3.3.1. Chemical and microbiological indicators of biodegradation 60
      3.3.2. Long-term dynamics of biochemical gradients 64
      3.3.3. Collapse of toluene degradation 66
      3.3.4. Short-term plume dynamics 68
      3.3.5. Transient hydraulic conditions and its potential influence on NA 69
   3.4. References 73

4. **Quantification and preservation of aquifer sediment bacteria – a multiple assay comparison 77**

   4.1. Introduction 78
   4.2. Materials and methods 80
      4.2.1. Sediment samples 80
      4.2.2. Culture bacteria 80
      4.2.3. Densitiy gradient centrifugation (DGC) 80
      4.2.4. Flow cytometry 81
      4.2.5. Microscopic counts 82
      4.2.6. Sediment preservation tests 83
      4.2.7. Detachment, loss and damage of cells 83
      4.2.8. Staining of prokaryotes 84
      4.2.9. Adenosine-triphosphate (ATP) 85
      4.2.10. Nucleic acid extraction and quantitative PCR 85
      4.2.11. Statistics 86
   4.3. Results and discussion 86
      4.3.1. Density gradient separation of bacteria and inorganic particles 87
      4.3.2. Evaluation of different fixatives 90
      4.3.3. Efficiency of cell dislodgement and recovery 91
      4.3.4. Comparison of different staining solutions 93
      4.3.5. Microscopy vs. flow cytometry 94
      4.3.6. Evaluated „standard protocol" 95
      4.3.7. Influence of sample storage 95
   4.4. Conclusion 98
   4.5. References 99

5. **General discussion and conclusion 106**

Appendix I

Authorship clarifications XV

Publications XVI

Acknowledgement ....

Abbreviations

A. aromaticum	*Aromatoleum aromaticum* EbN1
ATP	adenosine-5'-triphosphate
bls	below land surface
bss	benzyl succinate synthase
BTEX	benzene, toluene, ethylbenzene and the three isomeres of xylene
C-MLW	conventional multi-level well
CSIA	compound specific isotope analysis
DCG	density gradient centrifugation
DNA	deoxyribonucleic acid
DNAPL	dense non-aqueous phase liquids
dwt	dry weight
ENA	enhanceed natural attenuation
FSC	forward scatter
GC-MS	gas chromatography mass spectrometry
GC-C-IRMS	gas chromatography-combustion-isotope ratio mass spectrometry
HR-MLW	high-resolution multi-level well
IC_{50}	half maximal inhibiting concentration
K_f	hydraulic conductivity, permeability of the aquifer
LNAPL	light non-aqueous phase liquids
MQ	Milli-Q water
MUF-Glc	4-methylumbelliferyl β-D-glucoside
MUF-P	4-methylumbelliferyl phosphate
NA	natural attenuation
NAPL	non-aqueous phase liquids
PAH	polycyclic aromatic hydrocarbons
PCA	principal component analysis
PCR	polymerase chain reaction
PPI	sodium pyrophosphate
P. putida	*Pseudomonas putida* F1
qPCR	quantitative polymerase chain reaction
SIP	stable isotope probing
SSC	side scatter
TOM	total organic matter
T-RFLP	terminal restriction fragment length polymorphism
US EPA	United States Environmental Protection Agency

w/v	weight/volume
wwt	wet weight

1. Introduction

1.1. Groundwater – a vulnerable resource

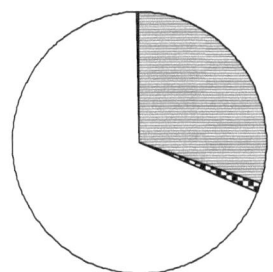

Figure 1.1: Global freshwater resources (Bannik et al. 2008).

The majority of global water resources consists of salt water, only 2.5% are fresh water. Out of that almost 70% is stored in ice, and rivers and lakes are less than 1%. Groundwater accounts for about 30% (Bannik et al. 2008) (Fig. 1.1) and, thus, it is an essential component of the hydrological cycle (Danielopol et al. 2003). Because of its natural abundance, while groundwater is the "key resource" for drinking water, only a small part is produced from surface and spring water. Other sources, such as desalinated sea water, are of negligible relevance. In Germany, for instance, more than 70% of the drinking water originates from groundwater (Bannik et al. 2008). Groundwater is generally of higher quality compared to surface water due to its purification via filtration through soil and sediments. In many countries clean and always available drinking water is a luxury, thus the demand for access to clean water was included to the human rights by the United Nations in July 2010 (UN 2010). In the future, water consumption will increase dramatically along with the growing world population, in particular because more and more water will be used for agriculture and industry. The intensive use and subsequent dramatic reduction of groundwater reservoirs will have a strong impact on the ecosystem. When groundwater is consumed faster than it is being replenished, groundwater tables are steadily falling. As the refill of groundwater reservoirs can take a very long time and fossil water resources are not renewable in the time span of human life, the renewal capacity of aquifers is a basic factor. Moreover, contamination as well as overdraft of water resources causes major problems to human and environmental health.

A serious problem for ecosystem groundwater is posed by anthropological pollution. Important sources of contaminations are oil and gasoline spills, as well as fertilizers and pesticides, landfills, urban and industrial wastewater, road salt or hazardous waste sites.

More than 322,000 sites are suspected to be contaminated in Germany alone, 25,000 of which are remediated, 4,000 are still in remediation and 3,500 are continuously monitored (LABO 2010).

The EU-Water Framework Directive (EU 2000) contains essential prerequisites for the protection of Europe's water, namely the Groundwater Daughter Directive which came into action in 2006 (EU 2006). Qualitatively good and "bad" groundwater is distinguished with the help of threshold values. An aquifer is in a good status when no salt or other intrusions exists, ecological or chemical quality of connected surface water is not at risk, dependent land ecosystems are not marred significantly and quality standards as well as threshold values were not exceeded at any of the monitoring wells at the aquifer (EU 2006). Aquifers of bad status must be improved with the aim to reach a good quality until 2015.

This thesis will focus on the fate of contamination of a sandy aquifer at a former gasworks site. Major pollutants are mono-aromatic hydrocarbons such as benzene, toluene, ethylbenzene and the three isomers of xylene (summarized as BTEX) as well as polyaromatic hydrocarbons (PAH) represented by naphthalene, acenaphthene and fluorene. Aromatic hydrocarbons are some of the most common organic contaminants in soil and groundwater and pose an eminent threat to our environment, due to their water solubility and toxicity (Kong & Johnstone 1994, Nahar et al. 2000, Röling & van Verseveld 2002, Kao et al. 2006).

1.2. Toxicity

Toxicity of hydrocarbons to living organisms is generated by accumulation in as well as disruption of cell membranes (Isken & de Bont 1998, Engraff et al. 2011) and can have various effects. BTEX compounds, for instance, exhibit acute toxicity after long-term exposure to animals and humans (Salantiro et al. 1997, ATSDR 2004, GESTIS 2010), which may result in damage of the central nervous system, also to the offspring (ATSDR 2004), and genotoxicity to lymphocytes (Chen et al. 2007). Furthermore, BTEX compounds in soil are toxic to plants (Salantiro et al. 1997, Henner et al. 1999, An 2004) as well as to bacteria (Salantiro et al. 1997, Dawson et al. 2008). Bacterial cells degrading these contaminants are more resistant, but nevertheless at higher concentrations these substances are also toxic for them. For instance, ethylbenzene concentrations >1 mM showed an inhibiting effect on biodegradation (Bauer 2007) and, more precise, a IC_{50} (half maximal inhibiting concentration) for *Pseudomonas* sp. for ethylbenzene, toluene and benzene of 0.8 mM, 2.9 mM and 4.4 mM, respectively, could be determined (Mattison et al. 2005).

The big group of PAH compounds can vary from nontoxic to extremely toxic and, thus, the United States Environmental Protection Agency (US EPA) listed 16 "priority PAHs" that are analyzed routinely for evaluation of contaminations worldwide (EPA 1998, Bojes & Pope 2007, UBA 2011). These contaminants can be carcinogenic, mutagenic and teratogenic (birth defect) to humans and animals (Cerniglia 1993, Engraff et al. 2011). At the site investigated in this study only naphthalene, acenaphthene and fluorene belonging to the 16 EPA-PAHs could be detected.

In addition, intermediates formed during contaminant degradation can be toxic, too, or even more toxic than their parent compounds. For instance, vinyl chloride from degradation of trichloroethylene (Wiedemeier et al. 1999), as well as sulfide, which is produced during sulfate reduction, are likely to inhibit biodegradation (van Leerdam et al. 2006). In addition to PAHs and metals, these intermediates furthermore increase toxicity in sediments (Tabak et al. 2003).

1.3. Transport and distribution

After release to the environment, petroleum hydrocarbons, such as BTEX and PAHs, generally seep through soil and unsaturated sediments into the underground. The heavier fraction of tar oils generally migrates downwards due to higher density forming so-called dense non-aqueous phase liquids (DNAPLs). Lighter compounds (Tab. 1.1), however, such as the mono-aromatic hydrocarbons and some low molecular weight PAHs, float as light non-aqueous phase liquids (LNAPLs) on the groundwater surface. Based on its higher water solubility this light fraction is successively dissolved and transported with the groundwater generating contaminant plumes.

Table 1.1: Physical-chemical properties of the most abundant contaminants detected at the Düsseldorf-Flingern site (modified from GESTIS 2010).

Compound	Structure	Molecular mass	Density at 20 °C	Solubility in water at 20 °C
Benzene		78.11	0.88 g cm^{-3}	1.77 g L^{-1}
Toluene		92.14	0.87 g cm^{-3}	0.47 g L^{-1}
Ethylbenzene		106.17	0.87 g cm^{-3}	0.14 g L^{-1}
o-Xylene		106.17	0.88 g cm^{-3}	0.18 g L^{-1}
m-Xylene		106.17	0.86 g cm^{-3}	0.2 g L^{-1}
p-Xylene		106.17	0.86 g cm^{-3}	0.2 g L^{-1}
Naphthalene		128.17	1.14 g cm^{-3}	0.032 g L^{-1}
Acenaphthene		154.21	1.22 g cm^{-3}	0.004 g L^{-1}

A contaminant plume spreading with the groundwater flow establishes a more or less constant dimension or, ideally, slowly shrinks in size (Wiedemeier et al 1999), so the risk for human and environmental health is minimized. The size and shape of the plume is controlled by complex interactions between biogeochemical and hydrogeological processes including biodegradation, diffusion, dispersion, dilution, sorption, ion exchange, precipitation, and change in recharge conditions (Christensen et al. 2001, Kneeshaw et al. 2007) (Fig. 1.2). Contaminants in groundwater moving through soil and sediment particles can be partly sorbed onto grain surfaces. Thus, the transport of contaminants can be decelerated or stopped. Additionally, many petroleum hydrocarbons evaporate readily into the atmosphere and thereby contaminant concentration in groundwater is reduced.

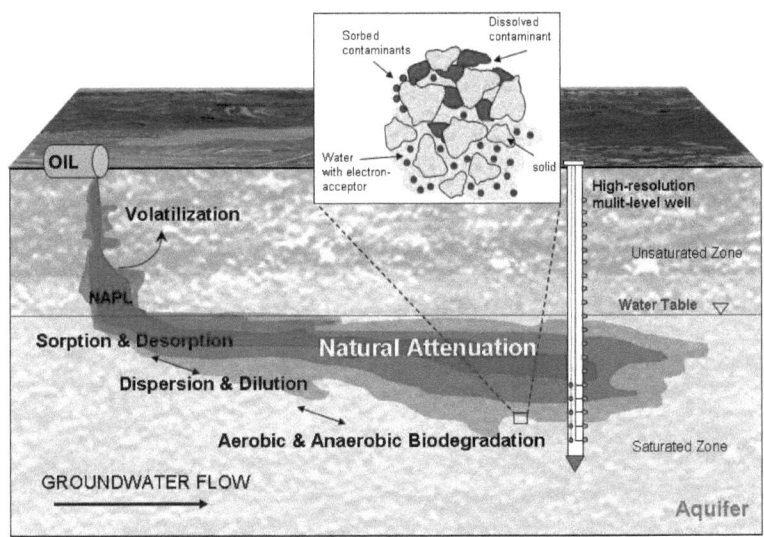

Figure 1.2: Transport and distribution of hydrocarbon contamination into the soil and groundwater.

The transport of substances across physical-chemical boundaries by mixing processes is regarded as the rate-controlling mechanism for many biogeochemical reactions (Thornton et al. 2001b, Klenk & Grathwohl 2002, Prommer et al. 2006, Maier et al. 2007). Transverse mixing regulates the transport of contaminants as well as of electron acceptors in the saturated zone. Diffusion and dispersion are the components of transverse mixing processes, where molecular diffusion is dominant only at very slow groundwater flow velocities (Grathwohl & Klenk 2000). As shown in experimental setups and reactive transport models small transverse dispersion coefficients in the range of millimeters to centimeters substantially restrict the lateral and vertical spreading of an organic contaminant plume (Ham et al. 2004, Cirpka et al. 2006, Maier et al. 2007, Bauer et al. 2008). Even though small-scale sediment heterogeneities or transient hydraulic conditions have been shown to be of minor relevance regarding the transverse dispersion coefficient, they potentially cause significant vertical plume meandering (Rahman et al. 2005). Hence, increasing convection is expected to be accompanied by an extension of the mixing area and, consequently, an increase in the scale of gradients.

The fringe of a contaminant plume constitutes its most important part, biodegradation is significantly increased in this area (Christensen et al. 2001, Mayer et al. 2001, Spence et al. 2001, Thornton et al. 2001a, Thornton et al. 2001b, Williams et al. 2001, Kneeshaw et al. 2007, Vencelides et al. 2007). Generally, metabolic activities in these zones are reflected by steep physical-chemical and microbial gradients confirming the "plume fringe concept": electron donors (the organic contaminants) and electron acceptors from the adjacent groundwater (e.g. sulfate) merge via dispersion (Fig. 1.2). An important additional reason for high activities in this area can be the dilution of contaminants to levels below the toxic concentration. Because redox conditions are dependent on solubility, mobility and bioavailability of mainly electron acceptors and contaminants (Christensen et al. 2001), they are not static in space and time (Skubal et al. 1999, Atteia & Guillot 2007, Vencelides et al. 2007).

1.4. Natural attenuation

Since pollution has become a worldwide problem, the remediation of contaminated groundwater is essential to ensure the quality of water resources, reduce human health risk and protect our environment. Natural attenuation (NA) combines physical, chemical and biological processes (Thornton et al. 2001a) to reduce mass, toxicity, mobility, volume and concentration of contaminants in soil and groundwater without human intervention (Christensen et al. 2001). NA is a cost-efficient, non-invasive treatment, but a time-intensive strategy to manage contaminated sites (Maier & Grathwohl 2006). Therefore, the investigation of parameters that affect the potential of NA is required. Biodegradation is the only process which leads to a mass reduction and thus the most important process (Cozzarelli et al. 1994) leading to an – ideally complete – removal of organic compounds in groundwater. Most of petroleum hydrocarbon contaminants are metabolized by microorganisms and biodegrading microbes can be found almost everywhere.

The OSWER (Office of Solid Waste and Emergency Remediation) directive of the US EPA suggests three lines of evidence that can be used to prove natural processes are active in reduction of contaminant concentrations. (1) Historical and monitoring data indicate a decrease of contaminant concentrations in association with a stable or retreating plume. (2) Geochemical indicators indirectly prove NA processes. During biodegradation background concentrations of electron acceptors (e.g. oxygen, nitrate, Fe(III) and sulfate) decrease and products such as dissolved iron(II) or sulfide are produced. (3) It need to be demonstrated that inherent bacteria degrade contaminants of concern either *in situ* with analysis of stable isotope data or in microcosms studies inoculated with material from the site (EPA 1999, 2010).

The NA approach is affected by some uncertainities such as possible resuspension of the contaminants as well as transport and distribution over wide areas (Tabak et al. 2003). Furthermore, NA alone can be insufficient to clean up a contaminated site in a reasonable time span. A combination of NA and active remediation may be applied to reduce the contaminant load more rapidly in such cases. The most common active remediation strategies are source reduction (EPA 2010) and pump-and-treat technologies. Due to their high financial costs they are not applied everywhere. It needs to be mentioned that NA processes may be sensitive to natural or anthropogenic changes of environmental conditions such as groundwater flow as well as to potential new and additional spills and release of contaminants into the subsurface.

1.5. Contaminant reduction – microbes as key players

Due to the low water solubility of oxygen as well as slow subsequent supply and replenishment into the groundwater along with simultaneous fast consumption by biotic and abiotic oxidation processes at organically contaminated sites, aerobic biodegradation is mainly limited to the area of the capillary fringe, the groundwater table and down-gradient of the source. Consequently, anaerobic reactions prevail in the major part of a contaminated aquifer. Towards the contamination source and the center of the plume, a sequence of different electron accepting conditions may be found. In general, the conditions are highly reduced close to the source and inside the plume (Christensen et al. 2001, Atteia & Guillot 2007). Degradation of petroleum hydrocarbons via oxygen reduction provides the highest energy yield to microbes (Tab. 1.2). Once oxygen is depleted, nitrate is reduced (Vencelides et al. 2007). Subsequently, in theory, manganese reduction follows combined with iron reduction, often overlapping with sulfate reduction as the further redox reaction and methanogenesis which is energetically less favorable (Skubal et al. 2001, Thornton et al. 2001b, Baez-Cazull et al. 2007, Vencelides et al. 2007). However, this classical redox zonation is not found at every contaminated aquifer as well as our test site (e.g. Lerner et al. 2000, Anneser et al 2008). Sulfate reduction seems to be the dominant anaerobic process in many organically polluted aquifers (Skubal et al. 1999, Wiedemeier et al. 1999, Thornton et al. 2001b, Kleikemper et al. 2002, Van Stempvoort et al. 2007). This is also the case at our test site (Anneser et al. 2008). Hence, sulfate reducing bacteria play a key role in anaerobic degradation (Van Hamme et al. 2003), mainly because of high sulfate solubility in combination with its natural occurrence in Germany's groundwater (Bannik et al. 2008).

Table 1.2: Terminal electron accepting processes (TEAPs) as exemplified for toluene oxidation (modified from Heider et al. 1999, Spence et al. 2005) with the overall free energy gained at pH 7 ($\Delta G°$).

Microbial process	Degradation reactions for toluene (C_7H_8)	$\Delta G°$ [kJ] per mol toluene
Aerobic degradation	$C_7H_8 + 9\ O_2 => 7\ CO_2 + 4\ H_2O$	-3913
Denitrification	$C_7H_8 + 7.2\ NO_3^- + 0.2\ H^+ => 7\ HCO_3^- + 0.6\ H_2O + 3.6\ N_2$	-3554
Mn(IV) reduction	$C_7H_8 + 18\ MnO_2 + 29\ H^+ => 7\ HCO_3^- + 18\ Mn^{2+} + 15\ H_2O$	-3502
Fe(III) reduction	$C_7H_8 + 94\ Fe(OH)_3 => 7\ FeCO_3 + 29\ Fe_3O_4 + 145\ H_2O$	-3398
Sulfate reduction	$C_7H_8 + 4.5\ SO_4^{2-} + 3\ H_2O => 7\ HCO_3^- + 2.5\ H_2S + 2\ HS^-$	-205
Methanogenesis	$C_7H_8 + 7.5\ H_2O => 2.5\ HCO_3^- + 2.5\ H^+ + 4.5\ CH_4$	-131

(Decreasing energy yield ↓)

Bacteria in aquifers have been shown to be the key players of groundwater quality, as they are mainly responsible for the degradation of organic pollutants (Chapelle 2001, Christensen et al. 2001). They provide the skills to transfer a wide range of contaminants (e.g. tar oil compounds and pesticides) under a great variety of environmental conditions (Heider et al. 1999, Widdel & Rabus 2001, Chakraborty & Coates 2004). In aerobic environments, virtually almost all organic compounds can be degraded by diverse metabolic pathways (Atlas 1981, Chapelle 2001). However, under anaerobic conditions numerous compounds, mainly the higher molecular compounds, are recalcitrant and often show a high persistence in nature (Philp et al. 2005). In case of the BTEX, all compounds can be degraded anaerobically with sulfate (Lovley et al. 1995, Lovley 1997, Kniemeyer et al. 2003, Morasch et al. 2004, Winderl et al. 2007). The absence or depletion of electron acceptors in groundwater plays an important role in BTEX-concentrations and the abundance of other organic compounds. Mass transfer of e.g. electron-acceptors or nutrients can limit the biodegradation capacity of microorganisms (Thornton et al. 2001b). Hydrocarbons which are dissolved in water, volatilized in the unsaturated zone or sorbed on soil particles are the most available to biodegradation by microorganisms. Bioavailability of all involved components (contaminant, electron-acceptor, nutrients) is crucial (Bosma et al. 1997, Tabak et al. 2003). Only molecules in their direct adjacencies can be utilized by microorganisms.

Microbial communities involved in contaminant degradation in the field and factors which control and limit their activities are still poorly understood. This is further hampered by the poor accessibility of the subsurface and the fact that merely ~1% of the inherent microbial community can be cultured in the laboratory (Amann et al. 1995, Watanabe 2001, O'Donnell et al. 2007). Fortunately, molecular techniques can circumvent this limitation, making the characterization of involved microorganisms without cultivation possible (Watanabe 2001). With these predominantly DNA-based assays it is possible to gain information on the abundance, diversity and activity of microbes in environmental samples (Amann et al. 1997, Lovley 2003). Combined with highly sensitive analytical techniques, a detailed description of the distribution of contaminants and redox-species is possible, as well as an assessment of complex interactions of biotic and abiotic processes taking place in contaminated groundwater and sediments. Furthermore, the localization of specific degraders within different redox zones of a contaminant plume allows the identification of degradation processes. It is important to consider that the bulk of bacteria is attached to the sediment and only a small amount is occurring suspended in the groundwater. However, subsurface sediment is difficult to be sampled. This is also true for the further analyses of sediment samples.

1.6. Evidence of biodegradation

Measurement of *in situ*-biodegradation is not a simple task. One frequently used tool is compound specific isotope analysis (CSIA) tackling contaminants (^{13}C), electron acceptors (e.g. ^{15}N in nitrate, ^{34}S and ^{18}O in sulfate) and reaction products (e.g. ^{34}S in sulfide) (Bolliger et al. 2001, Kleikemper et al. 2002, Richnow et al. 2003, Atekwana et al. 2005, Spence et al. 2005). Changes in stable isotope ratios in the residual fraction of the investigated compounds are almost exclusively a result of biological decomposition. Abiotic reactions are rarely responsible for these pronounced isotope effects (Ahad et al. 2000). However, isotope fractionation can be masked by the compound bioavailability and the bulk substrate concentration (Thullner et al. 2008). Furthermore, in case of some isotopes (e.g. ^{13}C/^{12}C) biodegradation must exceed to a high degree for detection of a distinct and pronounced stable isotope shift in the residual fraction of contaminant (Ahad et al. 2000, Meckenstock et al. 2004). The principle behind the observed isotope effects can be explained as follows: Elements can exist in more than one stable isotope, where the lighter one is naturally the more abundant. Enzymes discriminate between the heavy and light isotopes in bonds that are cleaved by processing molecules containing the lighter isotope at the specific cleavage site faster. This is because the required activation energy for the reaction is slightly lower. Therefore the lighter isotope becomes concentrated in the reaction products, while the heavier isotope accumulates in the remaining reactants. For example, strong sulfur isotope fractionation with enrichment of ^{34}S in sulfate is observed along with bacterial sulfate reduction (Bolliger et al. 2001, Kleikemper et al. 2004, Brunner et al. 2005, Van Stempvoort et al. 2007). Besides the sulfur isotopes, oxygen isotopes in sulfate may also provide a useful marker for different sources of sulfate (Spence et al. 2001). Likewise, enrichment of ^{13}C in toluene indicates biotic degradation of toluene.

The key enzyme for toluene as well as for xylene degradation is the benzylsuccinate synthase (bss), which catalyzes the fumarate addition reaction. The detection of the bss gene on DNA basis represents a qualitative hint for toluene degradation. However, it is not a tool for the quantification of biodegradation (Winderl et al. 2007). Here, the CSIA approach can help (e.g. Meckenstock et al. 2004).

Besides measuring specific activity, determining bacterial carbon production (BCP) or ATP content can give evidence about general microbial activity. BCP can be measured by means of incorporation of a labeled amino acid, e.g. tritium [^{3}H]-labeled leucine. The more active cells are, the more radioactive molecules are incorporated into their biomass, which can be quantified to assess the amount of BCP. ATP, which is a universal cell component – it is the energy currency – and is contained in living cells at a fairly constant amount, is depleted rapidly upon cell death. Therefore, ATP can be used as a basis for the estimation of viable microbial biomass from concentrations determined in water and sediment (Eydal & Pedersen 2007, Hammes et al. 2010).

1.7. High resolution sampling at a gasworks site

Field investigations during the thesis have been conducted at a former gasworks site located in the valley of the River Rhine in Düsseldorf-Flingern, Germany. The contamination from tar oil deposits were distributed in the shallow quaternary aquifer. This homogenous sandy and gravely zone was characterized by a mean K_f-value of 10^{-3} m s^{-1} and a velocity of 0.5 - 2 m day^{-1} (Anneser et al. 2008, Werner et al. 2008). The groundwater in this area is characterized by high sulfate concentrations, which are mainly caused by construction waste (Schmitt et al. 1998).

Until 1986 gas, tar oil, benzene, coke and fuels were produced at an area of about 170.000 m^2 (Wisotzky & Eckert 1997), polluting the soil over a period of more than 30 years. Consequently, a contaminant plume formed in the direction of the groundwater flow reaching a maximum dimension of 600 m in length and 100 m in width. Since 1995 the flowing off groundwater was remediated by "pump-and-treat". Additionally, the contaminant source could be reduced by excavation of contaminated soil in the area of the benzene plant in 1996. Thus, extension of the plume was reduced to a length of 100 m and a width of 50 m (Werner et al. 2008). Upgradient infiltration of high concentrations of nitrate in May 2003 as an enhanced natural attenuation (ENA) method increased the biological degradation potential. However, this treatment was aborted in December 2004 due to the insufficient reduction of the contaminant concentration (Richters 2008). The hydraulic remediation, however, reduced the length of the plume significantly, without impairing the time span of the remediation process (Werner et al. 2008). Therefore seven remediation wells were positioned in two plunger blocks where the contaminants never reached the second one. The contaminated groundwater after pumping is purified in a bioreactor, followed by filtration through gravel and activated charcoal filters. Afterwards, the treated water is introduced upstream of the contamination source via an infiltration ditch. Two of these pumping wells were deactivated in May 2003, but the remaining five wells are still working (Richters 2008). At the site a total of 70 groundwater monitoring wells (Werner et al. 2008) are installed and enable a horizontally and vertically resolved monitoring of the groundwater quality. The main contaminants present in the sandy aquifer are toluene, naphthalene and acenaphthene (KORA-Standortkompendium 2005, Anneser et al. 2008, Richters 2008). Naphthalene and acenaphthene are usually the most dominant contaminants in PAH-plumes, especially because of their relatively high solubility (Tab. 1.1) and low sorption capacity (Teutsch et al. 1997, Stupp & Paus 1999).

In order to investigate small scale substrate gradients, the presence of different redox zones and microbial patterns, a high-resolution multi-level well (HR-MLW) was developed and installed at the test site during the course of a former BMBF project in June 2005 in a distance of about 15 m to the contamination source (Anneser et al. 2008) (Fig. 1.3). Groundwater samples were collected simultaneously from various depths with a vertical resolution of up to 3 cm at a very low pumping rate of 1.5 mL min^{-1}, thus preventing the mixing of water from different depths. The HR-MLW is unique in its resolution and there are only a few other installations with similar high resolution of 10 - 30 cm (Schreiber et al. 2004) and 15 - 50 cm intervals (Griffioen 2001, van Breukelen et al. 2004, Vencelides et al. 2007).

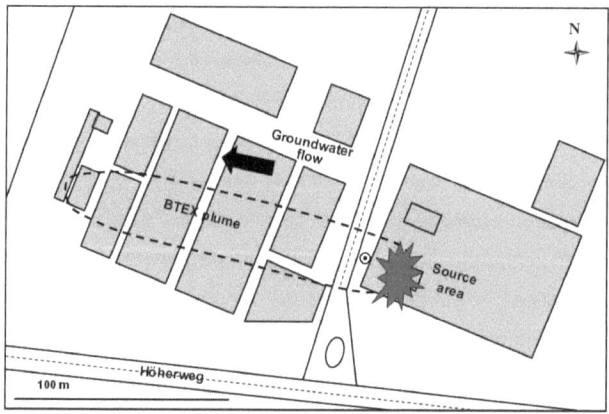

Figure 1.3: Schematic view of the site in Düsseldorf-Flingern with an assumption of the extension of the BTEX-plume and the location of the high-resolution well (◉).

The study of Anneser et al. (2008) showed that the contaminant plume at point of investigation is considerably thin, reaching a vertical expansion of 1 to 1.5 m only. Very steep and small-scale gradients define the plume fringes. Physical-chemical and stable isotope analysis point that sulfate reduction to be the most dominant redox process. Hot spots for biodegradation activity of toluene and xylene could be additionally confirmed with metabolites occurring mainly at the plume fringes (Jobelius et al. 2011).

1.8. Dynamics in the environment influencing bacterial communities and biodegradation

Shallow groundwater systems do not show the stability which was often assumed and proposed. Interestingly, during the observations, conducted by Bettina Anneser (2008) the measurement of small scale gradients revealed unexpected dynamics in the vertical dimension of the BTEX-plume showing pronounced changes within months, something uncoupled from pronounced changes in the groundwater table. For example, a substantial drop of the water table occurring in autumn 2006 was accompanied by a concomitant shift of the vertical localization of the BTEX-plume and redox gradients. The changed redox conditions show the adaption potential of the inherent microbial degrader community. As the sampling area is comprehensively sealed, the oscillations of the water table could not be directly correlated to precipitation data. Thus, it is speculative if the migration of the gradients is correlated to seasonal fluctuations. Rising water tables not only may cause a dilution and, consequently, toxicity reduction of the contaminants, may at worst float part of the contaminant source or new sources which have been in the unsaturated zone before.

Dynamic environmental conditions are known to influence microbial communities, whether predictable changes like in seasonally influenced tidal ecosystems or unexpected events like heavy rainfall on dry soil or disposal of contaminants. Dependent on seasonal parameters, e.g. temperature and wind induced mixing, of stratified lakes and, hence, the community composition and bacterial activity can shift within very short periods (Kisand & Noges 2004, Pearce 2005). Temperature, as a seasonally fluctuating parameter, causes also bacterial dynamics in the sea (Janelidze et al. 2011). Moreover, in highly dynamic environments, like estuarine systems, the microbial community is stronger influenced by spatial gradients (of e.g. salinity) than by seasonal variation (Lozupone & Knight 2007). Stable environments are associated with a greater diversity than disturbed conditions, like recharged-influenced environments in a contaminated aquifer (Haak et al. 2004).

Soil bacterial communities are also affected by seasonal variations as the supposable main factor influencing community pattern, biomass and activity (Blume et al. 2002, Hamel et al. 2006, Prevost-Boure et al. 2011). In a certain range, bacterial communities are very dynamic and microorganisms are adapted to different conditions as well as to fluctuating redox conditions (Hamel et al. 2006, De Angelis et al. 2010). Sudden dramatic events, however, can have a greater effect to active soil biomass than seasonal variations (Hamel et al. 2006). Consequences of occurring dynamics in an ecosystem are dependent on the resilience of this system. Holling described "ecosystem resilience" as the capacity of a natural system to cope with disturbances and restore its functional state (Holling 1973). Hence, biodiversity plays a crucial role by providing functional redundancy (Folke et al. 2004) and making it possible for ecosystems to reorganize after disturbances. Ecosystems with high resilience can tolerate disturbances without a collapse of the system. Groundwater is less affected by seasonal variations than surface water (Hancock et al. 2005), however, conditions can change dramatically when the unsaturated zone gets saturated or an uncontaminated area gets heavily contaminated. To date we do not know in which time frame these fluctuations do occur and how fast bacteria can cope with changed conditions.

1.9. Motivation and outline of the thesis

The present PhD thesis focuses on the spatial and temporal dynamics of selected redox processes and microbial parameters in a tar oil contaminated aquifer, aiming to gain deeper insights into limitations of NA. The prediction of time scales in which these dynamics occur and how microorganisms cope with changing conditions are tackled. Furthermore, I want to know to what extent it is possible to do general statements about NA events on basis of only water and sediment samples, and if only one type of sample can be enough.

Biodegradation of organic pollutants in porous groundwater systems may be limited by a number of factors such as the depletion of essential nutrients, the availability of electron acceptors, kinetic limitations, toxicity or simply the low number of microbial degraders (e.g. Lerner et al. 2000, Mayer et al. 2001, Bauer et al. 2008). Recent research suggested that the main limiting factor in sediment systems seems to be an insufficient mixing of electron donors (the contaminants) and electron acceptors. In the following, I wanted to know, if transient conditions enhance the NA capacity. Or do transient conditions impair the NA capacity? On the one hand, NA capacity could be enhanced by improved distribution and mixing of electron donors and -acceptors due to transient conditions. On the other hand, these dynamics could impair NA because established degrader communities cannot relocalize or readily react to changes. Dynamics of redox zones result in different electron accepting processes (Vencelides et al. 2007) as well as bacterial community changes and requires an adjustment of single species and groups to the altered conditions. Thus, one important investigative aspect is how fast microorganisms can respond to short-term and long-term changes of redox conditions and contaminant concentrations. Lag-phases in biodegradation due to environmental changes have been observed in many studies (Skubal et al. 2001, Thornton et al. 2001b, Kneeshaw et al. 2007). Long lag-phases in microbial activity may also occur because of toxicity of highly concentrated contaminants (Skubal et al. 1999), a crucial factor at a polluted site.

Sediment heterogeneity and transient conditions were shown to influence the extent of biodegradation (Bauer et al. 2008, Bauer et al. 2009), but which other factors affect NA capacity is not known. Additionally, the impacts of spatial and temporal variability are still far from fully understood (Fredrickson & Fletcher 2001). No general concept of the impact of fluctuations on the allocation of redox conditions and on biodegradation capacity exists.

Above all, the investigated gradients demonstrate the complexity of biodegradation processes, which were only accessible using a spatially and temporally adjusted sampling method. Thereby, it was possible to demonstrate the unexpected dynamics inherent in groundwater ecosystems, so far considered to be comparably stable environments. Availability of empirical data on small-scale spatial and temporal transient conditions was rare, even though they are found crucial for reliable assessment of NA and prediction calculations of site-specific reactive transport contamination. An improvement in predicting the plume movement is important to gain better acceptance of NA as a cost-efficient alternative to technical remediation strategies.

Moreover, a method for direct cell enumeration of bacteria attached to sediment particles was improved. The majority of bacteria live at and around sediment particles, not free-living in the groundwater. Thus, the cells have to be released from the sediment without destruction, followed by the separation of bacteria from interfering particles.

1.10. References

Ahad JME, Lollar BS, Edwards EA, Slater G, Sleep BE (2000) Carbon Isotope Fractionation during Anaerobic Biodegradation of Toluene: Implications for Intrinsic Bioremediation. Environmental Science and Technology 34:892-896

Amann R, Glockner FO, Neef A (1997) Modern methods in subsurface microbiology: in situ identification of microorganisms with nucleic acid probes. FEMS Microbiology Reviews 20:191-200

Amann R, Ludwig W, Schleifer K-H (1995) Phylogenetic Identification and In Situ Detection of Individual Microbial Cells without Cultivation. Microbiological Reviews 59:134-169

An Y-J (2004) Toxicity of Benzene, Toluene, Ethylbenzene, and Xylene (BTEX) Mixtures to *Sorghum bicolor* and *Cucumis sativus*. Bulletin of Environmental Contamination and Toxicology 72:1006-1011

Anneser B (2008) Spatial and temporal dynamics of biochemical gradients in a tar oil-contaminated porous aquifer – biodegradation processes revealed by high-resolution measurements. Eberhard-Karls-Universität, Tübingen

Anneser B, Einsiedl F, Meckenstock RU, Richters L, Wisotzky F, Griebler C (2008) High-resolution monitoring of biochemical gradients in a tar oil-contaminated aquifer. Applied Geochemistry 23:1715-1730

Atekwana EA, Atekwana E, Legall FD, Krishnamurthy RV (2005) Biodegradation and mineral weathering controls on bulk electrical conductivity in a shallow hydrocarbon contaminated aquifer. Journal of Contaminant Hydrology 80:149-167

Atlas RM (1981) Microbial Degradation of Petroleum Hydrocarbons: an Environmental Perspective. Microbiological Reviews 45:180-209

ATSDR (2004) Interaction Profile for: Benzene, Toluene, Ethylbenzene, and Xylenes (BTEX) U.S. Department of Health and Human Services. Agency for Toxic Substances and Disease Registry, Public Health Service, Atlanta, GA

Atteia O, Guillot C (2007) Factors controlling BTEX and chlorinated solvents plume length under natural attenuation conditions. Journal of Contaminant Hydrology 90:81-104

Baez-Cazull S, McGuire JT, Cozzarelli IM, Raymond A, Welsh L (2007) Centimeter-scale characterization of biogeochemical gradients at a wetland-aquifer interface using capillary electrophoresis. Applied Geochemistry 22:2664-2683

Bannik C, Engelmann B, Fendler R, Frauenstein J, Ginzky H, Hornemann C, Ilvonen O, Kirschbaum B, Penn-Bressel G, Rechenberg J, Richter S, Roy L, Wolter R (2008) Grundwasser in Deutschland. In: Bundesministerium für Umwelt NuRB (ed) Reihe Umweltpolitik, Berlin

Bauer RD (2007) Control and limitations of microbial degradation in aromatic hydrocarbon plumes – experiments in 2-D model aquifers. Eberhard-Karls-Universität, Tübingen

Bauer RD, Maloszewski P, Zhang Y, Meckenstock RU, Griebler C (2008) Mixing-controlled biodegradation in a toluene plume – Results from two-dimensional laboratory experiments. Journal of Contaminant Hydrology 96:150-168

Bauer RD, Rolle M, Bauer S, Eberhardt C, Grathwohl P, Kolditz O, Meckenstock RU, Griebler C (2009) Enhanced biodegradation by hydraulic heterogeneities in petroleum hydrocarbon plumes. Journal of Contaminant Hydrology 105:56-68

Blume E, Bischoff M, Reichert JM, Moorman TB, Konopka A, Turco RF (2002) Surface and subsurface microbial biomass, community structure and metabolic activity as a function of soil depth and season. Applied Soil Ecology 20:171-181

Bojes HK, Pope PG (2007) Characterization of EPA's 16 priority pollutant polycyclic aromatic hydrocarbons (PAHs) in tank bottom solids and associated contaminated soils at oil exploration and production sites in Texas. Regulatory Toxicology and Pharmaology 47:288-295

Bolliger C, Schroth MH, Bernasconi SM, Kleikemper J, Zeyer J (2001) Sulfur isotope fractionation during microbial sulfate reduction by toluene-degrading bacteria. Geochimica Et Cosmochimica Acta 65:3289-3298

Bosma TNP, Middeldorp PJM, Schraa G, Zehnder AJB (1997) Mass Transfer Limitation of Biotransforrmation: Quantifying Bioavailability. Environmental Science and Technology 31:248-252

Brunner B, Bernasconi SM, Kleikemper J, Schroth MH (2005) A model for oxygen and sulfur isotope fractionation in sulfate during bacterial sulfate reduction processes. Geochimica Et Cosmochimica Acta 69:4773-4785

Cerniglia CE (1993) Biodegradation of polycyclic aromatic hydrocarbons. Current Opinion in Biotechnology 4:331-338

Chakraborty R, Coates J (2004) Anaerobic degradation of monoaromatic hydrocarbons. Applied and Environmental Microbiology 64:437-446

Chapelle FH (2001) Groundwater Microbiology and Geochemistry. John Wiley & Sons, New York

Chen CS, Hseu YC, Liang CH, Kuo J-Y, Chen SC (2007) Assessment of genotoxicity of methyl-*tert*-butyl ether, benzene, toluene, ethylbenzene, and xylene to human lymphocytes using comet assay. Journal of Hazardous Materials 153:351-356

Christensen TH, Kjeldsen P, Bjerg PL, Jensen DL, Christensen JB, Baun A, Albrechtsen H-J, Heron G (2001) Biochemistry of landfill leachate plumes. Applied Geochemistry 16:659-718

Cirpka OA, Olsson A, Ju Q, Rahman MA, Grathwohl P (2006) Determination of transverse dispersion coefficients from reactive plume length. Groundwater 44:212-221

Cozzarelli IM, Baedecker MJ, Eganhouse RP, Goerlitz DF (1994) The geochemical evolution of low-molecular-weight organic acids derived from the degradation of petroleum contaminants in groundwater. Geochemica et Cosmochimica Acta 58:863-877

Danielopol DL, Griebler C, Gunatilaka A, Notenboom J (2003) Present state and future prospects for groundwater ecosystems. Environmental Conservation 30:104-130

Dawson JJC, Iroegbu CO, Maciel H, Paton GI (2008) Application of luminescent biosensors for monitoring the degradation and toxicity of BTEX compounds in soils. Journal of Applied Microbiology 104:141-151

De Angelis DF, Silver WL, Thompson AW, Firestone MK (2010) Microbial communities acclimate to recurring changes in soil redox potential status. Environmental Microbiology 12:3137-3149

Engraff M, Solere C, Smith KEC, Mayer P, Dahllöf I (2011) Aquatic toxicity of PAHs and PAH mixtures at saturation to benthic amphipods: Linking toxic effects to chemical activity. Aquatic Toxicology 102:142-149

EPA US (1998) Locating and estimating air emissions from sources of polycyclic organic matter. Office of Air Quality Planning And Standards Research Triangle Park

EPA US (1999) Use of monitored Natural Attenuation at Superfund, RCRA Corrective Action, and Underground Storage Tank Sites. OSWER directive 9200.4-17P, Washington, DC

EPA US (2010) Commonly asked questions regarding the use of natural attenuation for petroleum contaminated sites at federal facilities. In: www.epa.gov/swerffrr/documents/petrol.htm; accessed: 11.01.2011

EU (2000) Directive 2000/60/EC of the European Parliament and of the Council of 23 October 2000 establishing a framework for Community action in the field of water policy

EU (2006) Richtlinie 2006/118/EG des europäischen Parlaments und des Rates vom 12. Dezember 2006 zum Schutz des Grundwassers vor Verschmutzung und Verschlechterung

Eydal HSC, Pedersen K (2007) Use of an ATP assay to determine viable microbial biomass in Fennoscandian Shield groundwater from depths of 3-1000 m. Journal of Microbiological Methods 70:363-373

Folke C, Carpenter S, Walker B, Scheffer M, Elmqvist T, Gunderson L, Holling CS (2004) Regime Shifts, Resilience, and Biodiversity in Ecosystem Management. Annual Review of Ecology and Systematics 35:557-581

Fredrickson JK, Fletcher M (2001) Subsurface Microbiology and Biochemistryedn. Wiley-Liss, New York

GESTIS (2010) Stoffdatenbank www.dguv.de/ifa/stoffdatenbank. IFA Institut für Arbeitsschutz der Deutschen Gesetzlichen Unfallversicherung; accssed: 24.03.2011

Grathwohl P, Klenk I (2000) Transverse dispersion in aquifers: contaminant transport across the capillary fringe – mixing of electron acceptors and donors in plumes. In: Tratnyek PG, Adriaens P, Roden EE (eds) Chemical-Biological Interactions in Contaminant Fate. American Chemical Society, Washington, DC, p 369-371

Griffioen J (2001) Potassium adsorption ratios as an indicator for the fate of agricultural potassium in groundwater. Journal of Hydrology 254:244-254

Haak SK, Fogarty IR, West TG, Alm EW, McGuire JT, Long DT, Hyndman DW, Forney LJ (2004) Spatial and temporal changes in microbial community structure associated with recharge-influenced chemical gradients in a contaminated aquifer. Environmental Microbiology 6:438-448

Ham PAS, Schotting RJ, Prommer H, Davis GB (2004) Effects of hydrodynamic dispersion on plume lenghts for instantaneous bimolecular reactions. Advances in Water Resources 27:803-813

Hamel C, Hanson K, Selles F, Cruz AF, Lemke R, McConkey B, Zentner R (2006) Seasonal and long-term resource-related variations in soil microbial communities in wheat-based rotations of the Canadian prairie. Soil Biology & Biochemistry 38:2104-2116

Hammes F, Goldschmidt F, Vital M, Wang Y, Egli T (2010) Measurement and interpretation of microbial adenosine tri-phosphate (ATP) in aquatic environments. Water Research 44:3915-3923

Hancock PJ, Boulton AJ, Humphreys WF (2005) Aquifers and hyporheic zones: Towards an ecological understanding of groundwater. Hydrogeology Journal 13:98-111

Heider J, Spormann AM, Beller HR, Widdel F (1999) Anaerobic bacterial metabolism of hydrocarbons. FEMS Microbiology Ecology 22:459-473

Henner P, Schiavon M, Druelle V, Lichtfouse E (1999) Phytotoxicity of ancient gaswork soils. Effect of polycyclic aromatic hydrocarbons (PAHs) on plant germination. Organic Geochemistry 30:963-969

Holling CS (1973) Resilience and Stability of Ecological Systems. Annual Review of Ecology and Systematics 4:1-23

Isken S, de Bont JAM (1998) Bacteria tolerant to organic solvents. Extremophiles 2:229-238

Janelidze N, Jaiani E, Lashkhi N, Tskhvediani A, Kokashvili T, Gvarishvili T, Jgenti D, Mikashavidze E, Diasamidze R, Narodny S, Obiso R, Whitehouse CA, Huq A, Tediashvili M (2011) Microbial water quality of the Georgian coastal zone of the Black Sea. Marine Pollution Bulletin 62:573-580

Jobelius C, Ruth B, Griebler C, Meckenstock RU, Hollender J, Reineke A, Frimmel FH, Zwiener C (2011) Metabolites Indicate Hot Spots of Biodegradation and Biochemical Gradients in a High-Resolution Monitoring Well. Environmental Science & Technology 45:474-481

Kao CM, Huang WY, Chang LJ, Chen TY, Chien HY, Hou F (2006) Application of monitored natural attenuation to remediate a petroleum-hydrocarbon spill site. Water Science and Technology 53:321-328

Kisand V, Noges T (2004) Abiotic and biotic factors regulating dynamics of bacterioplankton in a large shallow lake. FEMS Microbiology Ecology 50:51-62

Kleikemper J, Schroth MH, Bernasconi SM, Brunner B, Zeyer J (2004) Sulfur isotope fractionation during growth of sulfate-reducing bacteria on various carbon sources. Geochimica et Cosmochimica Acta 68:4891-4904

Kleikemper J, Schroth MH, Sigler WV, Schmucki M, Bernasconi SM, Zeyer J (2002) Activity and diversity of sulfate-reducing bacteria in a petroleum hydrocarbon-contaminated aquifer. Applied and Environmental Microbiology 68:1516-1523

Klenk ID, Grathwohl P (2002) Transverse vertical dispersion in groundwater and the capillary fringe. Journal of Contaminant Hydrology 58:111-128

Kneeshaw TA, McGuire JT, Smith EW, Cozzarelli IM (2007) Evaluation of sulfate reduction at experimentally induced mixing interfaces using small-scale push-pull tests in an aquifer-wetland system. Applied Geochemistry 22:2618-2629

Kniemeyer O, Fischer T, Wilkes H, Glöckner FO, Widdel F (2003) Anaerobic Degradation of Ethylbenzene by a New Type of Marine Sulfate-Reducing Bacterium. Applied and Environmental Microbiology 69:760-768

Kong S, Johnstone D (1994) Toxicity of toluene and o-xylene to *Acinetobaceter calcoaceticus* in starvation-survival mode. Biotechnology Letters 16:1217-1220

KORA-Standortkompendium (2005) Kontrollierter natürlicher Rückhalt und Abbau von Schadstoffen bei der Sanierung kontaminierter Grundwässer und Böden.

LABO (2010) Bundesweite Kennzahlen zur Altlastenstatistik. In: Altlastenstatistik_Juli_2010_d09

Lerner DN, Thornton SF, Spence MJ, Banwart SA, Bottrell SH, Higgo JJ, Mallison HEH, Pickup RW, Williams GM (2000) Ineffective Natural Attenuation of Degradable Organic Compounds in a Phenol-Contaminated Aquifer. Groundwater 38:922-928

Lovley DR (1997) Potential for anaerobic bioremediation of BTEX in petroleum-contaminated aquifers. Journal of Industrial Microbiology & Biotechnology 18:75-81

Lovley DR (2003) Cleaning up with Genomics: Applying Molecular Biology to Bioremediation. Nature Reviews Microbiology 1:35-44

Lovley DR, Coates J, Woodward JC, Phillips EJP (1995) Benzene Oxidation Coupled to Sulfate Reduction. Applied and Environmental Microbiology 61:953-958

Lozupone CA, Knight R (2007) Global patterns in bacterial diversity. PNAS 104:11436-11440

Maier U, Grathwohl P (2006) Numerical experiments and field results on the size of steady state plumes. Journal of Contaminant Hydrology 85:33-52

Maier U, Rügner H, Grathwohl P (2007) Gradients controlling natural attenuation of ammonium. Applied Geochemistry 22:2606-2617

Mattison RG, Taki H, Harayama S (2005) The Soil Flagellate *Heteromita globusa* Accelerates Bacterial Degradation of Alkylbenzenes through Grazing and Acetate Excretion in Batch Culture. Microbial Ecology 49:142-150

Mayer KU, Benner SG, Frind EO, Thornton SF, Lerner DN (2001) Reactive transport modeling of processes controlling the distribution and natural attenuation of phenolic compounds in a deep sandstone aquifer. Journal of Contaminant Hydrology 53:341-368

Meckenstock RU, Morasch B, Griebler C, Richnow HH (2004) Stable isotope fractionation analysis as a tool to monitor biodegradation in contaminated aquifers. Journal of Contaminant Hydrology 75:215-255

Morasch B, Schink B, Tebbe CC, Meckenstock RU (2004) Degradation of *o*-xylene and *m*-xylene by a novel sulfate-reducer belonging to the genus *Desulfotomaculum*. Arch Microbiol 181:407-417

Nahar N, Alauddin M, Quilty B (2000) Toxic effects of toluene on the growth of activated sludge bacteria. World Journal of Microbiology and Biotechnology 16:307-311

O'Donnell AG, Young IM, Rushton SP, Shirley MD, Crawford JW (2007) Visualization, modelling and prediction in soil microbiology. Nature Reviews Microbiology 5:689-699

Pearce DA (2005) The structure and stability of the bacterioplankton community in Antarctic freshwater lakes, subject to extremely rapid environmental change. FEMS Microbiology Ecology 53:61-72

Philp J, Bamforth SM, Singleton I, Atlas RM (2005) Environmental pollution and restoration: a role for bioremediation. In: Atlas RM, Philp J (eds) Bioremediation. ASM Press, Washington, DC

Prevost-Boure NC, Maron PA, Ranjard L, Nowak V, Dufrene E, Damesin C, Soudani K, Lata J-C (2011) Seasonal dynamics of the bacterial community in forest soils under different quantities of leaf litter. Applied Soil Ecology 47:14-23

Prommer H, Tuxen N, Bjerg PL (2006) Fringe-controlled natural attenuation of phenoxy acids in a landfill plume: Integration of field-scale processes by reactive transport modeling. Environmental Science & Technology 40:4732-4738

Rahman MA, Jose SC, Nowak W, Cirpka OA (2005) Experiments on vertical transverse mixing in a large-scale heterogeneous model aquifer. Journal of Contaminant Hydrology 80:130-148

Richnow HH, Meckenstock RU, Reitzel LA, Baun A, Ledin A, Christensen TH (2003) In situ biodegradation determined by carbon isotope fractionation of aromatic hydrocarbons in an anaerobic landfill leachate plume (Vejen, Denmark). Journal of Contaminant Hydrology 64:59-72

Richters L (2008) Grundwassersanierung, Stadtwerke Düsseldorf, Düsseldorf

Röling WFM, van Verseveld HW (2002) Natural attenuation: What does the subsurface have in store? Biodegradation 13:53-64

Salantiro JP, Dorn PB, Huesemann MH, Moore KO, Rhodes IA, Rice Jackson LM, Vipond TE, Western MM, Wisnieswski HL (1997) Crude Oil Hydrocarbon Bioremediation and Soil Ecotoxicity Assessment. Environmental Science & Technology 31:1769-1776

Schmitt R, Langguth H-R, Püttmann W (1998) Abbau aromatischer Kohlenwasserstoffe und Metabolitenbildung im Grundwasserleiter eines ehemaligen Gaswerkstandorts. Grundwasser - Zeitschrift der Fachsektion Hydrogeologie 2:78-86

Skubal KL, Barcelona MJ, Adriaens P (2001) An assessment of natural biotransformation of petroleum hydrocarbons and chlorinated solvents at an aquifer plume transect. Journal of Contaminant Hydrology 49:151-169

Skubal KL, Haack SK, Forney LJ, Adriaens P (1999) Effects of dynamic redox zonation on the potential for natural attenuation of trichloroethylene at a fire-training-impacted aquifer. Physics and Chemistry of the Earth Part B - Hydrology Oceans and Atmosphere 24:517-527

Spence MJ, Bottrell SH, Thornton SF, Lerner DN (2001) Isotopic modelling of the significance of bacterial sulphate reduction for phenol attenuation in a contaminated aquifer. Journal of Contaminant Hydrology 53:285-304

Spence MJ, Bottrell SH, Thornton SF, Richnow HH, Spence KH (2005) Hydrochemical and isotopic effects associated with petroleum fuel biodegradation pathways in a chalk aquifer. Journal of Contaminant Hydrology 79:67-88

Stupp HD, Paus HL (1999) Migrationsverhalten organischer Grundwasser-Inhaltsstoffe und daraus resultierende Ansätze zur Beurteilung von Monitored Natural Attenuation (MNA). Terra Tech 5:1-14

Tabak HH, Lazorchak JM, Lei L, Khodadoust AP, Antia JE, Bagchi R, Suidan MT (2003) Studies on Bioremediation of polycyclic Aromatic Hydrocarbon-Contaminated Sediments: Bioavailability, Biodegradability, and Toxicity Issues. Environmental Toxicology and Chemistry 2:473-482

Teutsch G, Gratwohl P, Schiedik T (1997) Literaturstudie zum natürlichen Rückhalt / Abbau von Schadstoffen im Grundwasser. Handbuch Altlasten und Grundwasserschadensfälle. Landesanstalt für Umweltschutz, Karlsruhe

Thornton SF, Lerner DN, Banwart SA (2001a) Assessing the natural attenuation of organic contaminants in aquifers using plume-scale electron and carbon balances: model development with analysis of uncertainty and parameter sensitivity. Journal of Contaminant Hydrology 53:199-232

Thornton SF, Quigley S, Spence MJ, Banwart SA, Bottrell S, Lerner DN (2001b) Processes controlling the distribution and natural attenuation of dissolved phenolic compounds in a deep sandstone aquifer. Journal of Contaminant Hydrology 53:233-267

Thullner M, Kampara M, Richnow HH, Harms H, Wick LY (2008) Impact of Bioavailability Restrictions on Microbial Induced Stable Isotope Fractionation. 1. Theoretical Calculation. Environmental Science & Technology 42:6544-6551

UBA (2011) EPA-Liste www.umweltprobenbank.de/de/documents/13446, Dessau-Roßlau; accessed: 23.03.2011

UN (2010) The human right to water and sanitation. General Assembly

van Breukelen BM, Griffioen J, Röling WFM, van Verseveld HW (2004) Reactive transport modelling of biochemical processes and carbon isotope geochemistry inside a landfill leachate plume. Journal of Contaminant Hydrology 70:249-269

Van Hamme JD, Singh A, Ward OP (2003) Recent advances in petroleum microbiology. Microbiology and Molecular Biology Reviews 67:503-549

van Leerdam RC, de Bok FAM, Lomans BP, Stams AJM, Lens PNL, Jannsen AJH (2006) Volatile Organic Sulfur Compounds in Anaerobic Sludge and Sediments: Biodegradation and Toxicity. Environmental Toxicology and Chemistry 25:3101-3109

Van Stempvoort DR, Armstrong J, Mayer B (2007) Microbial reduction of sulfate injected to gas condensate plumes in cold groundwater. Journal of Contaminant Hydrology 92:184-207

Vencelides Z, Sracek O, Prommer H (2007) Modelling of iron cycling and its impact on the electron balance at a petroleum hydrocarbon contaminated site in Hnevice, Czech Republic. Journal of Contaminant Hydrology 89:270-294

Watanabe K (2001) Microorganisms relevant to bioremediation. Current Opinion in Biotechnology 12:237-241

Werner P, Börke P, Hüsers N (2008) Leitfaden Natürliche Schadstoffminderung bei Teerölaltlasten. In: KORA IfAuA (ed) Themenverbund 2: Gaswerke, Kokereien, Teerverarbeitung, (Holz)Imprägnierung. Technische Universität Dresden, Dresden

Widdel F, Rabus R (2001) Anaerobic biodegradation of saturated and aromatic hydrocarbons. Current Opinion in Biotechnology 12:259-276

Wiedemeier TH, Rifai HS, Newell CJ, Wilson JT (1999) Natural attenuation of fuel and chlorinated solvents in the subsurfaceedn. John Wiley & Sons, Chichester

Williams GM, Pickup RW, Thornton SF, Lerner DN, Mallison HEH, Moore Y, White C (2001) Biochemical characterisation of a coal tar distillate plume. Journal of Contaminant Hydrology 53:175-197

Winderl C, Schaefer S, Lueders T (2007) Detection of anaerobic toluene and hydrocarbon degraders in contaminated aquifers using benzylsuccinate synthase (bssA) genes as a functional marker. Environmental Microbiology 9:1035-1046

Wisotzky F, Eckert P (1997) Sulfat-dominierter BTEX-Abbau im Grundwasser eines ehemaligen Gaswerksstandortes. Grundwasser – Zeitschrift der Fachsektion Hydrogeologie 1/97:11-20

2. High resolution analysis of contaminated aquifer sediments and groundwater – what can be learned in terms of natural attenuation?

High-resolution depth-resolved monitoring was applied to groundwater and sediments samples in a tar oil contaminated aquifer. Today, it is not fully clear, whether groundwater-based lines of evidence are always sufficient to adequately assess natural attenuation (NA) potentials and processes going on *in situ*. Our data unveiled small-scale heterogeneities, steep physical-chemical and microbial gradients, as well as hot spots of contaminants and biodegradation in the supposedly homogeneous sandy aquifer. The comparison of basic geochemical data revealed a fairly good agreement between sediment and groundwater samples. Nevertheless, a comprehensive understanding of both BTEX and PAH distribution, as well as redox processes involving insoluble electron acceptors, *i.e.* iron reduction, clearly asks for consideration of both, sediment and groundwater analysis. A thin BTEX plume right below the groundwater table was visible only in groundwater, while significant amounts of PAHs were present in sediments from deeper zones of the aquifer. Indications for sulfate reduction as a dominant process involved in BTEX degradation were largely obtained from groundwater, while the role of iron reduction in degradation and possible sulfide oxidation at the capillary fringe and the upper BTEX plume fringe, as well as in deeper PAH-contaminated zones was evident from sediments. Moreover, sediment analysis were essential to meaningfully recover cellular abundances and distribution, activity and composition of the bacterial community.

Sediments harbored >97.7% of microbial cells and displayed enzyme activities 5 to 6 orders of magnitude higher than groundwater samples. Bacterial community T-RFLP fingerprints revealed important distinctions, but also similarities in depth-resolved microbial community distribution in sediments and water. An apparently highly specialized degrader population was found to dominate the lower BTEX plume fringe. However, even though sediment data seemed to comprise most community information found also in groundwater, this relation did not apply *vice versa*. In summary, our results show that groundwater sampled at an appropriate scale may contain sufficient information to identify and localize dominant redox reactions, but clearly fails to unravel natural attenuation potentials. This clearly emphasizes the importance of both groundwater and sediment samples for truly assessing natural attenuation potentials and activities at organically contaminated aquifers.

2.1. Introduction

Contaminations with complex mixtures of petroleum hydrocarbons are known to exhibit a long-term persistence in soils and aquifers, and may be detectable even after thousands of years (Eberhardt & Grathwohl 2002). In many cases, a complete remediation of contaminations fails, despite application of vigorous physical-technical remediation (e.g. air sparging, thermal treatment or pump-and-treat techniques) or biostimulation approaches. The monitoring of natural attenuation (NA) is therefore often chosen as a cost- and labour-efficient strategy to follow the transport and fate of pollutants in subsurface environments (Bamforth & Singleton 2005). Consequently, there is a growing interest to predict the spread and development of contaminant plumes and to understand intrinsic degradation behavior. NA is controlled by a complex interplay of physical, chemical and biological reactions, which are in turn affected by sediment-porewater interactions (McGuire et al. 2000). The distribution and transport of contaminants in groundwater, the distribution and structure of microbial degrader populations, as well as actual biodegradation activities are at least partially determined by sediment and groundwater characteristics.

The monitoring of NA at a given site usually involves the investigation of a limited set of groundwater parameters (*i.e.* contaminants, redox species), but rarely also considers geochemical and microbiological analyses of sediments, or both groundwater and sediments (Bekins et al. 2001, Lehman et al. 2001b). For aquifer microbial populations, however, it is known that the major part occurs associated to sediments (e.g. Hazen et al. 1991, Albrechtsen & Winding 1992, Alfreider et al. 1997, Griebler et al. 2002), and activities are often found higher for attached cells (e.g. Holm et al. 1992, Madsen & Ghiorse 1993, Albrechtsen & Christensen 1994). Previous studies have also revealed that considerable discrepancy in microbial community composition and biodegradation potential can prevail between sediment and groundwater samples of the same location

(Lehman et al. 2001b, Röling et al. 2001, Lehman & O'Connell 2002, Brad et al. 2008). It is therefore not fully clear, whether monitoring of groundwater is sufficient to adequately assess *in situ* NA processes and potentials at a specific site. It is frequently hypothesized that most degradation processes in aquifers are catalyzed by attached microbial populations (Kölbel-Boelke et al. 1989, Röling et al. 2001, Lehman & O'Connell 2002). However, respective data sets comparing NA process information obtainable in sediment vs. water samples taken at a given site are extremely rare. Due to the considerable efforts associated with borehole drilling and installation of multi-level monitoring wells, only a limited number of studies is available where both groundwater and sediment analysis have been conducted in parallel today. Furthermore, the spatial resolution of sampling may often not be sufficient to adequately recover small-scale sediment heterogeneities and the reactive biogeochemical gradient zones at the fringes of contaminant plumes (Cozzarelli et al. 1999, Kao et al. 2001, Anneser et al. 2008b, Brad et al. 2008, Winderl et al. 2008). In this study, we present a comprehensive data set comparing the biogeochemistry and microbiology of groundwater and sediment samples from a tar oil-contaminated sandy aquifer.

Temporally and spatially closely connected sampling allowed a direct comparison of both sediment and groundwater characteristics and enabled us to localize and scrutinize NA potentials. In particular, the distribution of contaminants and redox reaction products in groundwater and sediments was assessed in relation to the abundance, activity, diversity and distribution of bacteria present in different zones of the contaminated aquifer. Important distinctions in the two different data sets were more than evident. While distinct abiotic features may be sufficiently recovered by groundwater monitoring, this study especially points out the importance of sediment matrix-based information for assessing major microbial NA potentials.

2.2. Materials and methods

2.2.1. Site description

The investigated area is a Quaternary sandy aquifer at a former gasworks site in Düsseldorf-Flingern, Germany. Release of tar oil compounds during operation and breakdown of the plant resulted in the development of a contaminant plume with concentrations of up to 100 mg L^{-1} of monoaromatic and 10 mg L^{-1} of polycyclic aromatic hydrocarbons (PAHs) dissolved in groundwater. In the course of several remediation actions launched since 1995 the major part of the tar oil phases was removed. Nevertheless, residual amounts still can be detected in sediments and groundwater, forming a hydrocarbon plume of about 200 x 40 m in horizontal dimension. For detailed information on site characteristics see (Eckert 2001, Anneser et al. 2008a, Anneser et al. 2008b).

2.2.2. Groundwater sampling

In February 2006, groundwater was collected from a specially designed high-resolution multi-level well (HR-MLW), which extends from 3 m to 12 m below land surface (bls) and exhibits sampling filter spacings of 2.5 cm, 10 cm and 30 cm. For geochemical analyses, groundwater was sampled at vertical intervals of 2.5 - 10 cm from 6.38 m (groundwater table) down to 8.1 m bls in 100 mL pre-washed narrow neck glass bottles and immediately processed for the analysis of a number of biotic and abiotic parameters, as described below. Water samples for microbial community analyses were sampled from only 5 selected depths down to 9 m bls. Here, water was collected in autoclaved 1 L glass bottles, sealed on site to minimize oxygen exposure, and transported to the lab within 24 h under cooling (<10°C). In the lab, ~750 mL per depth were filtered over 0.22 µm cellulose acetate filters (Corning Inc., NY, USA), which were immediately frozen until DNA extraction. A detailed description of the HR-MLW as well as information on the sampling procedure and the materials and instruments used is given in (Anneser et al. 2008a, Anneser et al. 2008b).

2.2.3. Sediment sampling

One week after sampling groundwater, two boreholes were drilled in a distance of approximately 0.5 m downgradient and, respectively, 1.5 m adjacent to the HR-MLW by means of hollow-stem auger drilling. Sediment liners were obtained in one meter segments down to a depth of 14 m (Liner 1) and 11 m (Liner 2), respectively. Right after retrieving the sediment liners they were transferred into an aluminum box and subsampled immediately under argon atmosphere for a selected set of parameters. Sediment sub cores from selected depths were immediately fixed for individual parameter measurements (see below), or frozen on dry ice for subsequent DNA extraction and molecular community analyses (Winderl et al. 2008).

2.2.4. Sample preparation and geochemical analysis

Groundwater collected in 100 mL glass bottles was immediately processed for measuring specific conductivity, pH and redox potential as well as for the analysis of dissolved iron and sulfide species, which is described in detail in (Anneser et al. 2008a). Total reduced inorganic sulfur was extracted from sediment samples fixed in 20% (w/v) zinc acetate using HCl-Cr(II) solution (1 M Cr(II)-HCl in 12 M anoxic HCl), according to the protocol of (Ulrich et al. 1997). Under maintenance of anoxic conditions, reduced inorganic sulfur species were converted to hydrogen sulfide during 30 h of intensive shaking at room temperature. The liberated H_2S was trapped in a 10% zinc acetate solution and subsequently analyzed analogous to groundwater sulfide measurements (Anneser et al. 2008a).

Groundwater samples dedicated to the analysis of dissolved mono- and polycyclic aromatic hydrocarbons were amended with NaOH (0.1 M final concentration) to stop biological activity. BTEX concentrations were measured via GC-MS by headspace analysis; the less soluble polycyclic aromatic hydrocarbons (PAH) were determined by liquid injection analysis. Major anions (SO_4^{2-}, HPO_4^{3-}) and cations were analyzed via ion chromatography (Dionex DC-100, Idstein, Germany). Further details on analytical procedures are given in (Anneser et al. 2008a).

Sediment total organic matter (TOM) was calculated from the loss in weight of dried sediment after combustion of the organic content at 450°C for 4 h. Amorphous, ready available sedimentary iron was extracted in duplicates with 30 mL of 0.5 M HCl from 2 mL of sediment during 24 h of intensive shaking at room temperature (Heron et al. 1994). Likewise, crystalline iron phases were obtained by 5 M HCl extraction over a period of 21 days. The extracted Fe(II) was measured analogous to groundwater samples (Anneser et al. 2008a). Concentrations of ferric iron species were determined after reduction of total iron to Fe(II) using hydroxylamine hydrochloride (HACl; 10% w/w in 1 M HCl) as reductant. The amount of HACl-reducible Fe(III) was calculated as the difference of Fe(II) determined in the HCl and HACl extracts (Lovley & Phillips 1987, 1988).

PAHs adsorbed to the sediment were extracted with acetone and amended with an internal standard mixture containing deuterated acenaphthene, chrysene, perylene and phenanthrene species (Internal Standards Mix 25, Ehrenstorfer, Augsburg, Germany) at a final concentration of 25 mg L^{-1}. Aromatic hydrocarbons were determined in a GC-MS applying the settings as described in (Winderl et al. 2008).

In situ microbial activity was assessed by determining the enzymatic hydrolysis rates of two fluorogenic substrates, *i.e.* MUF-P (4-methylumbelliferyl phosphate) and MUF-Glc (4-methylumbelliferyl β-D-glucoside) (Freeman et al. 1995, Hendel & Marxsen 1997). Enzyme assays of both groundwater and sediment samples were performed in 4 mL Supelco glass vials. For each sample, three replicates and one control were prepared. Groundwater was filled immediately after sampling into the vials avoiding headspace. Sediment samples were transferred in aliquots of 1 mL into vials, which have been prefilled with anoxic, low-ionic strength (1:500 dilution) freshwater medium (Widdel & Bak 1992). The substrates MUF-P or MUF-Glc (both Sigma, Taufkirchen, Germany) were added to a final concentration of 225 µM to groundwater samples and 1 mM to sediment samples, respectively. After incubation at *in situ* temperature (16°C ± 1°C) for 18 h, groundwater samples were stopped by addition of ammonium glycine buffer (pH 10.5) at a ratio of 1:10. Enzyme activity within sediment samples was terminated after 22 h of incubation (16°C ± 1°C) by adding 100 µL of an alkaline solution (2 M NaOH / 0.4 M EDTA) to each sample. Controls for background fluorescence were treated analogously, but inactivated prior to incubation. Fluorescence was determined in a luminescence spectrometer (Aminco Bowman Series 2, Thermo Scientific) at 455 nm emission and 365 nm excitation.

2.2.5. Stable isotope analysis of sulfate

Stable sulfur ($^{34}S/^{32}S$) and oxygen ($^{18}O/^{16}O$) isotope ratios of sulfate dissolved in groundwater were determined from samples collected in 100 mL vessels containing 10 mL of a 20% (w/v) Zn-acetate solution for precipitation of sulfide. Procedures for the recovery of sulfate and details on isotope measurements are described elsewhere in detail (Anneser et al. 2008b).

2.2.6. Determination of bacterial cell numbers

From sediment liners stored at -20°C, 0.5 mL sample aliquots were fixed with 2.5% glutardialdehyde and kept at 4°C until further preparations. After replacement of the glutardialdehyde by 1.5 mL PBS, cells were released from sediment using a swing mill (Retsch, MM 200; 3 min, 20 Hz) and separated from abiotic particles via density gradient centrifugation according to the protocol of (Lindahl & Bakken 1995). The layer containing the bulk (about 80%) of bacterial cells was collected and cells were stained with SYBR Green I (Molecular Probes, Invitrogen, Karlsruhe, Germany) at a ratio of 1:10,000. Total cell counts were quantified in a flow cytometer (LSR II, Becton Dickinson, Heidelberg, Germany) equipped with a 488 nm and 633 nm laser, using Trucount beads (Trucount tubes, Becton Dickinson) as internal standard. Instrument settings were as follows: forward scatter (FSC) 350 mV, side scatter (SSC) 300 - 370 mV, B530 (bandpass filter 350 nm) 500 - 580 mV. All parameters were collected as logarithmic signals. For minimization of background noise, the threshold was adjusted to 200 mV each for FSC and SSC. Calculations of total bacterial cell numbers were performed as described in (Nebe-von-Caron et al. 2000). Loss of cells due to sediment freezing was accounted for applying a correction factor determined from fresh and frozen sediments from a later sampling at the same site.

2.2.7. Bacterial community fingerprinting

DNA was extracted from 9 different sediment depths of Liner 1 (6, 6.4, 6.65, 6.9, 7.1, 7.6, 7.8, 8.1 and 9.2 m) and available corresponding groundwater filtrates (6.38, 6.46, 6.83, 7.1, 8.1 and 9.0 m). The extraction protocols used for sediments (~1 g wwt) and filters were as previously described (Winderl et al. 2008, Brielmann et al. 2009), using a CTAB extraction buffer (Zhou et al. 1996) modified as follows: 100 mM Tris, 100 mM Na-EDTA, 95 mM Na_2HPO_4, 750 mM NaCl, 1% CTAB, pH 7.80. Extracted DNA was stored frozen (-20°C) until further analyses. Terminal restriction fragment length polymorphism (T-RFLP) analysis of bacterial 16S rRNA gene amplicons was done with primers Ba27f-FAM / 907r and *Msp*I digestion as previously described (Winderl et al. 2008). Data evaluation and principal component analyses (PCA) was performed as reported elsewhere (Lueders & Friedrich 2002, Winderl et al. 2008).

2.3. Results

2.3.1. Distribution of contaminants

The spread of major contaminants (e.g. benzene, toluene, ethylbenzene and xylenes; BTEX) in groundwater of the investigated aquifer section was restricted to a narrow zone stretching from right below the groundwater table (ca. 6.4 m bls at time of sampling) to a depth of approximately 8.1 m bls, describing a vertically thin plume with a maximum BTEX concentration of 58 mg L^{-1} at 6.7 m depth (Fig. 2.1A).

Toluene constituted the main fraction of monoaromatic hydrocarbons, accounting for about 75% of total BTEX concentrations. For naphthalene, the major dissolved PAH component, a similar depth profile was found with a maximum concentration of 16 mg L^{-1} at 6.75 m bls (Fig. 2.1A). Out of the 16 EPA-PAHs, only acenaphthene and fluorene were further detected in groundwater, with lowest concentrations in the area of the BTEX plume and slightly increased values (<1 mg L^{-1}) further down (>7 m bls) (data not shown).

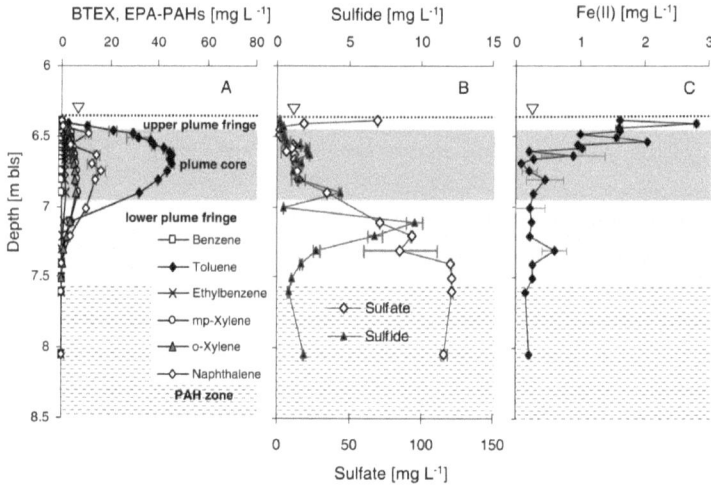

Figure 2.1: Small-scale distribution of (A) contaminants, (B) sulfate and sulfide, and (C) ferrous iron in groundwater sampled from the HR-MLW; values represent means of duplicate measurements ± SD.

Sediment samples contained no detectable amounts of BTEX, but exhibited distinct peaks for individual PAHs such as naphthalene, acenaphthene and fluorene. In detail, concentration maxima within Liner 1 were found at 6.65 m and 8.8 m bls (Fig. 2.2A). Both maxima were mainly contributed by naphthalene, with the upper peak located directly within the center of the plume, and the lower in an area where no BTEX and only minor concentrations of PAHs, *i.e.* acenaphthene and fluorene, could be detected in groundwater. Interestingly, no other of the 16 EPA-PAHs were found or, if yet, they were below the detection limits of the respective extraction and analysis protocols (*i.e.* 20 µg L^{-1} for groundwater and 6.8 µg kg^{-1} wwt for sediments). The distribution patterns of total PAHs within Liner 2 were found to be quite similar to Liner 1 for the upper layers of the saturated zone, yet concentrations of individual components partly differed. Additionally, a PAH-impacted layer that was not present in Liner 1 was detected at 10.2 m bls (Fig. 2.2B), thus pointing at lateral sediment heterogeneities.

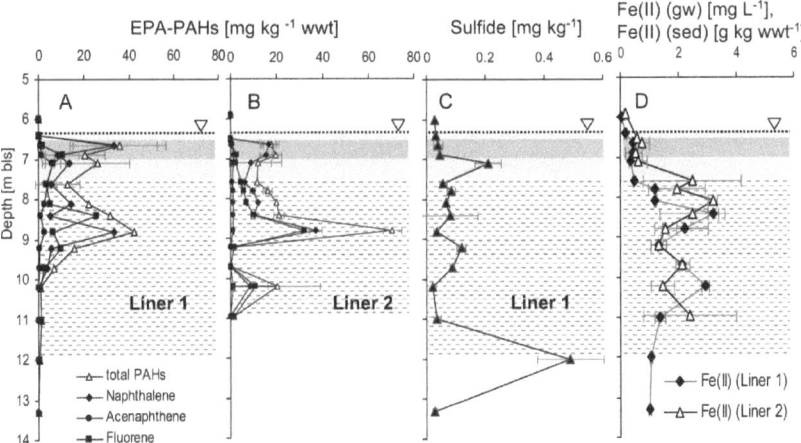

Figure 2.2: Vertical distribution of (A, B) contaminants, (C) sulfide, and (D) ferrous iron in sediments in relation to BTEX (grey area) and PAH (shaded area) contamination zones, respectively; values represent means of duplicate measurements ± SD.

2.3.2. Physical-chemical conditions and redox-specific parameters

The redox potential (E_h) in groundwater rapidly declined with depth from -20 mV at the uppermost sampling point to a value of -235 mV in the plume center (6.81 m; Tab. 2.1). A trend opposite to E_h was observed for pH, starting at a value of 7.07 right below groundwater table and a maximum of 7.46 at 6.81 m bls (Tab. 2.1). Likewise, specific conductivity ranging from 1030 to 1215 µS cm^{-1} showed an increasing trend with depth (Tab. 2.1). Dissolved inorganic phosphate was available all across the investigated aquifer section, with lowest concentrations of 70 µg L^{-1} at the water table and peak values of up to 2.1 mg L^{-1} in the plume center (Tab. 2.1). No nitrate (detection limit of 0.1 mg L^{-1}) could be detected in groundwater at time of sampling. Sulfate, the most important dissolved electron acceptor, was almost absent in the BTEX plume center, yet its availability increased rapidly with declining contaminant concentrations at the upper and lower plume fringe (Fig. 2.1B). At a depth of 7.4 m bls, sulfate leveled off to a concentration of about 120 mg L^{-1}.

Table 2.1: Hydrogeochemistry and stable isotope ratios of groundwater sampled from the HR-MLW.

Depth [m]	pH	Redox [mV]	Specific conductivity [µS cm^{-1}]	HPO_4^{2-} [mg L^{-1}]	SD	$\delta^{34}S$ [‰]	SD	$\delta^{18}O$ [‰]	SD
6.39	7.07	-20	1096						
6.41		-100	1030	0.07					
6.46	7.13	-100	1053	0.54	0.02				
6.56	7.27	-150	1052	2.05	0.18				
6.67	7.40	-190	1081	1.19	0.17				
6.75	7.43	-195	1094	1.07	0.38				
6.81	7.46	-235	1038	1.18		42.4	0.17	15.2	0.04
7.11	7.42	-230		0.44	0.05	25.0	0.20	12.6	0.13
7.41	7.39	-197	1215	0.95	0.12				
8.08	7.43	-229	1166	0.57	0.06	15.9	0.02	11.1	0.11
9.05						12.5	0.02	10.7	0.05

Stable isotope values of sulfate, *i.e.* $^{34}S/^{32}S$ and $^{18}O/^{16}O$, determined for selected depths indicated the occurrence of bacterial sulfate reduction in the contaminated aquifer

(Tab. 2.1). Concomitant with declining concentrations of organics and increasing sulfate concentrations at the lower plume fringe, distinct peaks of dissolved sulfide were observed, reaching a maximum of almost 10 mg L^{-1} at 7.11 m bls (Fig. 2.1B). Lower amounts of sulfide were detected in the plume core and scarcely any dissolved sulfide was found within the first centimeters beneath the capillary fringe. There, however, elevated concentrations of dissolved Fe(II) were measured, followed by a steep decline until approximately constant concentrations of about 0.3 mg L^{-1} were attained at 6.9 m bls (Fig. 2.1C).

In contrast, ferrous iron extracted from the sediments (0.5 M HCl) exhibited lowest concentrations within the BTEX contamination area but exhibited a pronounced increase with depth, reaching concentrations of up to 3.2 g kg^{-1} wwt beneath 8 m bls (Fig. 2.2D). With the exception of local peaks at 10.2 m (Liner 1) and 9.2 m (Liner 2) bls, the recovery of ready extractable Fe(III) from saturated sediments was considerably lower compared to that of Fe(II), i.e. always <1 g kg^{-1} in Liner 1 and, respectively, <2 g kg^{-1} wwt in Liner 2 (data not shown). Only low amounts of crystalline Fe(II) averaging 1 mg kg^{-1} were obtained by extraction with 5 M HCl; in contrast, concentrations of Fe(III) released by this harsh extraction exhibited clearly higher values of up to 5.5 g kg^{-1} wwt (data not shown). Total reduced inorganic sulfur extracted from sediment samples of Liner 1 revealed concentrations in the range of 0.02 - 0.5 g sulfide per kg (wwt), with pronounced peaks at 7.1 and 12 m depth (Fig. 2.2C). Unfortunately, no sulfide values are available from Liner 2 sediments. Total organic matter (TOM) content in dried sediments of both liners ranged between 0.2 and 0.6% (dwt), with a mean value of 0.34% in Liner 1 and 0.39% in Liner 2.

2.3.3. Direct cell counts and enzyme activities

In sediments of Liner 1, bacterial cell counts exhibited two distinct peaks of 4.1×10^7 and 6.8×10^7 cells g^{-1} wwt at the upper and lower fringe of the BTEX plume, separated by a local minimum of 8.8×10^6 cells g^{-1} wwt in the plume center (Fig. 2.3). Background numbers in the less contaminated zone below 8 m bls averaged at 6.8×10^6 cells g^{-1} wwt. In contrast, direct cell counts in sediments of Liner 2 exhibited one dominant peak of 4.8×10^7 cells g^{-1} wwt at the upper plume fringe (Fig. 2.3) and only a less distinct increase of cell number at the lower plume fringe. With depth, cell numbers decreased significantly down to background concentrations of 6.9 x 10⁶ cells g⁻¹ wwt below 8 m bls. Unfortunately, no bacterial cell numbers are available from groundwater from this sampling event.

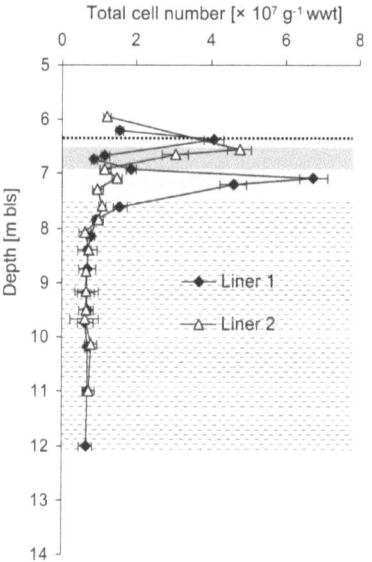

Figure 2.3: Total number of bacteria in sediments from Liner 1 and 2 in relation to BTEX- and PAH-contaminated zones; values represent means of triplicate measurements ± SD.

Both sediment and groundwater samples exhibited maximum rates of substrate conversion at the uppermost sampling depths. Up to 1.9 mmol MUF-P were hydrolyzed by phosphatases per kilogram sediment (wwt) and hour in Liner 2 (Fig. 2.4C). Turnover rates in Liner 1 reached maximum values of 1.2 mmol kg⁻¹ wwt h⁻¹ (Fig. 2.4B). At 6.9 m bls, the activity started to level off at 180 µmol kg⁻¹ wwt h⁻¹ (Liner 2) and 60 µmol kg⁻¹ wwt h⁻¹ (Liner 1) on average. The gradient of MUF-P in groundwater samples was less pronounced and the activities were significantly lower, showing values of max. 32 nmol L⁻¹ h⁻¹ directly below the water table (Fig. 2.4A), which corresponds to 10.9 nmol kg⁻¹ h⁻¹ wwt assuming a sediment porosity of 34%. Consequently, phosphatase activities in sediments were found up to six orders of magnitude higher than in groundwater. Sediment β-glucosidase activities were comparable for both liners according to the individual depth zones (Fig. 2.4B+C). Hydrolysis rates of MUF-Glc in groundwater exhibited slightly elevated values at the upper and lower fringe of the BTEX plume (Fig. 2.4A). Similar to MUF-P turnover, sediment activities outreached groundwater activities by five orders of magnitude.

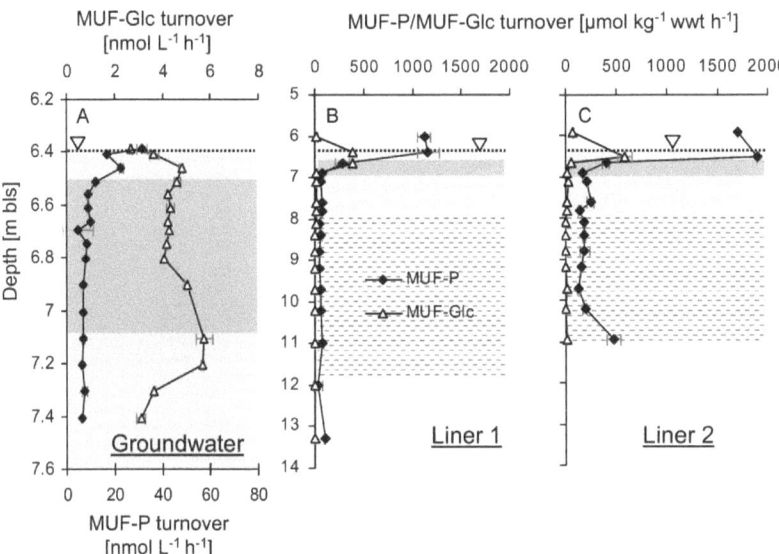

Figure 2.4: Microbial enzyme activities in (A) groundwater and sediments of (B) Liner 1 and (C) Liner 2 in relation to BTEX (grey) and PAH (shaded) contaminated zones; values represent means of triplicate measurements ± SD.

2.3.4. Depth-resolved analysis bacterial communities

T-RFLP analysis of bacterial communities indicated pronounced shifts in structure and diversity of total bacterial communities with depth (fingerprints are not shown). For groundwater, the Shannon-Wiener diversity H' as inferred from relative T-RF abundances was highest at the level of the capillary fringe/groundwater table and dropped to local minima within the lower plume fringe (7.1 m) (Fig. 2.5A). For sediments, H' was very similar with depth, except at the lower part of the plume core, where a slightly increased local maximum of bacterial diversity was observed. A total of 77 distinct T-RFs was retrieved from both sediment and water samples, of which 21 peaks were unique for the sediment and 17 for the water samples. However, with the exception of the uppermost samples from near the capillary fringe, H' was always lower in water samples than in corresponding sediments.

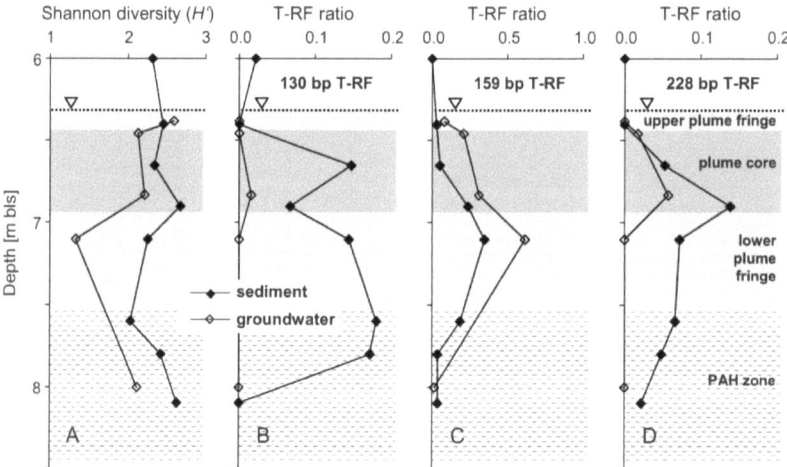

Figure 2.5: Depth-distribution of (A) bacterial T-RFLP fingerprinting Shannon-Wiener diversity H' and relative T-RF abundance of characteristic T-RFs; i.e. related to (B) *Geobacter* spp. (130 bp), (C) *Desulfocapsa* spp. (159 bp), and (D) clostridial sulfate reducers and fermenters (228 bp) in sediments and corresponding groundwater samples of the Flingern site. The putative affiliation of the T-RFs is in accordance to previously published clone libraries from the same site (Winderl et al. 2008).

To further elucidate these distinctions, we compared the depth-distribution of the relative abundance of selected T-RFs previously identified to represent important components of the degrader community. Respective results for sediment samples taken in 2005 revealed that especially the 130, 159, and 228 bp T-RFs represented dominating members of the contaminant-degrading bacterial community established in the lower plume fringe. As shown in a previous study by 16S rRNA gene cloning, these T-RFs were affiliated to microbes related to *Geobacter* spp., *Desulfocapsa* spp., and also clostridial sulfate reducers and fermenters (Winderl et al. 2008). Albeit semi-quantitative, this assessment revealed pronounced distinctions in distribution patterns (Fig. 2.5). T-RF abundances of over 15% were observed for 130 bp peaks (*Geobacter*-related microbes) in sediments of the plume core and over the lower plume fringe. In groundwater samples, however, this T-RF was only detected in one depth (6.83 m) and was thus almost not detectable over the entire depth transect (Fig. 2.5B). In contrast, the 159 bp T-RF (*Desulfocapsa*-related) showed a very similar distribution with depth in both sample sets, with clear abundance maxima in the lower plume fringe at 7.1 m (Fig. 2.5C). The 159 bp T-RF was always of a higher relative abundance in groundwater than in sediments (~62 vs. ~35%, respectively). Finally, the 228 bp T-RF (putatively representing clostridial sulphate reducers and fermenters) was present at a maximum frequency of ~14% and ~6% in both sediments and water of the lower plume core, respectively (Fig. 2.5D).

To comparatively assess total bacterial community variance over depth in both sample sets, PCA statistics of the T-RFLP data set was conducted. The percentage of total community variability explained by the two primary PC factors inferred was no less than 59% (Fig. 2.6). Surprisingly, data reduction indicated congruent community dynamics with depth over the different zones of the Flingern plume. As observed previously (Winderl et al. 2008), especially the populations of the lower fringe zone were resolved in ordination, indicating similar community patterns for both sediments and groundwater samples in these depths. As observed before for the loading of PCA factors on specific T-RFs (Winderl et al. 2008), the distinction of the lower plume fringe degrader community was largely attributed to a high abundance of the 159 bp T-RF (factor loadings not shown). In summary, PCA illustrated similar community dynamics (clockwise shifts in virtual ordination) with depth for both sample sets, but dynamics (*i.e.* distinctions) with depth were more pronounced for groundwater samples (Fig. 2.6).

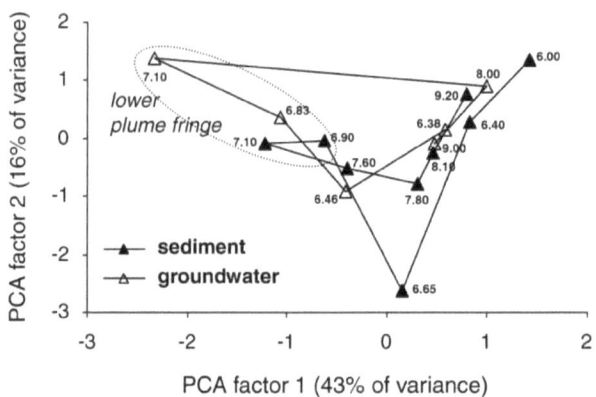

Figure 2.6: Ordination of PCA factor scores for T-RFLP fingerprinting of sediment and water samples comprising the overall variance in depth-resolved bacterial community composition. The depths at which specific fingerprints were retrieved are indicated next to the ordination points.

2.4. Discussion

Efficient biodegradation of organic contaminants in the subsurface requires two elements: (1) microbial populations with degradative capabilities, and (2) favourable geochemical and hydrological conditions (McGuire et al. 2000). Then, *vice versa*, biodegradation processes are suggested to cause significant changes in groundwater and sediment properties, which include: (i) a rapid depletion of the energetically most favorable electron acceptors (*i.e.* oxygen and nitrate) (Anderson & Lovley 1997, Christensen et al. 2000), (ii) the appearance of metabolites and reaction products (e.g. reduced sulfur and iron species as shown in Fig. 2.2C+D), (iii) increase in pH and alkalinity owing to proton consumption during anaerobic oxidation of organic compounds, and (iv) a decrease in redox potential as well as a change in specific conductivity due to the conversion of individual ions such as sulfate (Cozzarelli et al. 1999, Christensen et al. 2001). Some of these indicators may easily be followed in groundwater samples, however, some may clearly call for sediment data. For example, with respect to important redox processes (distribution of major electron acceptors and reaction products) information from groundwater samples is limited to dissolved chemical species. The role of solid electron acceptors such as ferric iron and the turnover of contaminants adsorbed to the sediment matrix can only be assessed directly in sediments, or indirectly via soluble metabolites and end products in groundwater. In contrast, major dissolved electron donors cannot be adequately monitored via sediment samples, and only partial information on intrinsic microbial communities and respective activities is obtainable from groundwater. As we discuss in the following, both, groundwater and sediments needs to be addressed when assessing NA potential in an organically polluted aquifer.

2.4.1. Small scale physical-chemical heterogeneities in groundwater and sediments

Steep concentration gradients in groundwater clearly delineated a fine-scale vertical zonation of the BTEX plume. As obvious from Fig. 2.1A, only the extraordinarily high-resolution of groundwater sampling allowed depicting the small-scale and steep geochemical gradients, which could not have been resolved by a conventional multi-level well with a vertical resolution of 0.5 to 1 m. However, groundwater samples did not provide comprehensive information regarding the PAH contamination. This is not surprising, as PAHs exhibit a high tendency to adsorb to the sediment matrix. A recent study (Paissé et al. 2008) found up to 90% of PAHs adsorbed and only a small fraction dissolved in groundwater. Naphthalene, due to its comparably low molecular weight and high solubility (Eberhardt & Grathwohl 2002), showed a similar distribution in groundwater as the monoaromatic hydrocarbons. Compared to the maximally 16 mg L^{-1} of naphthalene dissolved in groundwater (= 5.4 mg per kg soil; Tab. 2.2), maximally ~6 to 7-fold increased amounts were found sorbed to the sediments (Fig. 2.2A+B). Sorption of PAHs to the sediment matrix is substantially influenced by the presence of organic matter, which is commonly regarded to be negligible in sandy aquifers, where mean organic carbon contents are in the range of only about 0.1% (Lovley & Chapelle 1995, Appelo & Postma 2005). Nevertheless, due to their hydrophobicity, the higher molecular weight PAHs acenaphthene and fluorene are strongly retained. In the Flingern aquifer, despite the relatively low soil organic carbon content of ~0.3% (dwt), concentrations of acenaphthene were 10 to 20-fold, fluorene concentrations even 100-fold higher in sediments than in groundwater.

The relation between adsorbed (C_s) and dissolved concentrations (C_w) can be expressed by the solid water distribution coefficient K_D of a given compound, according to:

$$K_D = \frac{C_S \left[mg\ kg^{-1} \right]}{C_W \left[mg\ kg^{-1} \right]} \quad (1)$$

In soil-water systems, K_D is also defined by the product of the organic carbon-water distribution coefficient K_{oc} and the fraction f_{oc} of organic compounds in solution (Appelo & Postma 2005):

$$K_D = K_{oc} \times f_{oc} \quad (2)$$

Using literature K_{oc} values (Tab. 2.3) and measured f_{oc} (= 0.3%) and C_w in eq. 1 and 2, the theoretically adsorbed contaminant concentration C_s can be calculated (Tab. 2.2). Measured concentrations of fluorene and naphthalene adsorbed to the sediment matrix exceeded calculated results by a factor of 5 and 2.5, respectively.

Table 2.2: Sorption properties and aqueous solubilities S_i of selected aromatic hydrocarbons. [a](Eberhardt & Grathwohl 2002), [b](Appelo & Postma 2005), [c](USACE 2002) and [d](Szabo et al. 1990).

Compound	Log K_{ow}	Log K_{ow}	S_i at 25°C [mg L^{-1}]
Toluene	2.69[c]	2.18[d]	534.8
Naphthalene	3.36[b]	2.94[b]	31.7
Acenaphthene	3.98[c]	3.66[c]	3.93
Fluorene	4.18[c]	3.86[c]	1.98

S_i = aqueous solubility of organic compound i, K_{ow} = octanol-water coefficient, K_{ow} = organic carbon-water coefficient

Amounts of acenaphthene determined in the sediment agreed fairly well to theoretical values, overestimating the actual mean concentration by 10% (Tab. 2.2). Calculated maximum toluene concentrations adsorbed to the sediment matrix amounted to

~6.8 mg kg^{-1}, but neither toluene nor other BTEX compounds were detected in sediment samples of the Flingern site. Apparently, the sorption capacity of the sediment matrix for these substances was lower than expected. In conclusion, sediment samples alone would lead to inconclusive information on the BTEX contamination.

Table 2.3: Measured and calculated maximum concentrations of aromatic hydrocarbons in sediment and groundwater of the Flingern aquifer.

Compound	Groundwater [mg L^{-1}]	Liner 1[b]	Linder 2[b]	Calculated[c]
		Sediment [mg kg^{-1}]		
Toluene	45 (15.3)	-	-	6.80
Naphthalene	16 (5.4)	33.5	37	13.94
Acenaphthene	0.9 (0.31)	6	1.6	4.11
Fluorene	0.7 (0.24)	26	32	5.07

[a] parentheses indicate the corresponding sedimentary concentrations in mg kg^{-1}
[b] maximum concentrations measured in the respective sediment liners
[c] theorethically sorbed contaminant concentration calculated from max. contaminant concentrations measured in groundwater

Considering the inaccuracies associated with the collection of sediments by liner coring, the differences in the distribution patterns of contaminants may also partly be ascribed to on site heterogeneities. "Hot spots" of pollutants and preferential flow paths in heterogeneous sediments may foster a diffuse spreading of contaminants. In the source zone, which is located ~15 m upstream of the sampling area (Anneser et al. 2008b), the dense non-aqueous phase liquids (DNAPLs) are likely to seep down due to their higher density and molecular weight, which may explain the accumulation of PAHs in deeper zones in contrast to the lighter BTEX compounds.

2.4.2. Biodegradation via sulfate reduction

A significant enrichment of both ^{18}O and ^{34}S isotopes in groundwater sulfate proved microbial sulfate reduction to occur throughout the BTEX-contaminated zone, with pronounced activities across the lower plume fringe (Tab 2.1; see also Anneser et al.

2008a). Inverse gradients of dissolved sulfate and sulfide (Fig. 2.1B), which are regarded as indicators of ongoing sulfate reduction, support this hypothesis. Further geochemical evidence for bacterial sulfate reduction derived from a distinct peak of reduced inorganic sulfur extracted from sediment samples, which again hints at high sulfate reduction activity at 7.1 m bls, marking the transition between heavily and less contaminated groundwater in the lower plume fringe (Fig. 2.2C). The lack of sulfide at the upper plume fringe is most likely attributed to scavenging of sulfide by both complexation and reoxidation reactions with Fe(II) (Stumm & Morgan 1996, Cozzarelli et al. 1999, Kostka et al. 2002).

2.4.3. Iron reduction and iron cycling processes

Iron reduction has previously been shown to occur at the Flingern site (Eckert 2001). Groundwater analysis revealed iron reducing activity being most pronounced at the upper and lower plume fringe (Eckert 2001, Anneser et al. 2008a). Elevated concentrations of dissolved Fe(II) measured in groundwater samples hint at significant iron reduction in the upper 20 cm of the saturated zone (Fig. 2.1C). However, only sediment analysis could show that considerable amounts of readily extractable ferric iron are actually present in this zone (data not shown). At the interface between the unsaturated and saturated zone, cycling of iron species is likely to proceed at high rates. Transient hydraulic conditions causing a fluctuating groundwater table govern regular periods of re-oxidation and reduction, providing a regeneration of sedimentary electron acceptors (Ulrich et al. 1998, Ulrich et al. 2003). A minor secondary peak of ferrous iron in groundwater was observed at the lower plume fringe. This pattern has already been observed in previous investigations at that site (Anneser et al. 2008a, Anneser et al. 2008b).

From a thermodynamic point of view, iron reduction is considered to be energetically more favorable than sulfate reduction (Lovley & Chapelle 1995). Nevertheless, bacterial iron reduction in organically contaminated porous aquifers is frequently limited by the availability of Fe(III) species (Albrechtsen & Christensen 1994, Tuccillo et al. 1999). In the sediments of Liner 1 and 2, extractable ferric iron did only exceptionally account for more than 1 g kg^{-1} wwt. Crystalline Fe(III) species, on the other hand, were available at two to fourfold higher concentrations all across the anoxic aquifer section, with maxima above and below the BTEX plume (data not shown). Although bacteria are theoretically capable of using crystalline Fe(III) as electron acceptor (Roden & Zachara 1996), these are only slowly converted and less attractive to microorganisms owing to their stability and inaccessibility (Lovley 2001).

It has to be considered, furthermore, that only a small ratio of the ferrous iron is found dissolved in groundwater, while the major part remains associated with to sediment surfaces. This can further decrease the availability of sedimentary Fe(III) to attached bacteria. Interestingly, readily extractable ferrous iron in sediments exhibited highest values (3 - 4 g kg^{-1} wwt) below the BTEX plume (Fig. 2.2D). Sediment data thus points at an undepleted reservoir of ferric iron in deeper zones of the aquifer. Independent from depth the amount of Fe(II) associated with the sediment in the saturated zone accounted for >99%, while less than 1% was found dissolved in groundwater. A similar distribution of Fe(II) between groundwater and sediment matrix was also reported for other aquifers (Lovley & Phillips 1988, Heron et al. 1994). Elevated concentrations of Fe(II) and concurrent low sulfide concentrations, as found near the capillary fringe, point also at sulfide oxidation with ferric iron, a topic which needs further investigations.

2.4.4. Microbial patterns in groundwater and sediments

It has repeatedly been shown that cells attached to surfaces can account for 90 to 99.99% of the microbial biomass in porous aquifers (Griebler & Lueders 2009). The sediment cell counts, therefore, matched well general findings. However, the two biomass peaks in Liner 1 and 2 (here the lower one was less pronounced) impressively mirrored zones of highest biodegradation activities as indicated by redox gradients. Unfortunately, no data on the number of suspended cells were available for the February 2006 sampling campaign due to accidental loss of sample vials.

Thus, no direct sediment-groundwater comparisons could be performed for this particular parameter. However, the sediment data obtained may be well compared with groundwater data from a sampling campaign six months later (Anneser et al. 2008b). The general distribution of bacteria in groundwater was highly similar to the sediment pattern reported here. In detail, the total number of cells was highest in the uppermost sampling point in the upper plume fringe zone, exhibiting 1.2×10^6 cells mL^{-1}. In the plume core, cell numbers were ten times less with an average value of 2.1×10^5 mL^{-1}, but increased again to 5 x 10^5 mL^{-1} at the lower plume fringe. Further down, a slight decrease of cell numbers to 3.2 x 10^5 mL^{-1} was observed, interrupted by another distinct peak (6.8 x 10^5 mL^{-1}) right in the area of highest PAH concentrations in the sediment (Anneser et al. 2008b). Comparing total cell counts in groundwater and sediments on a volume basis (assuming a sediment active porosity of 34%) revealed between 97.7 and 99.8% of the cells attached to the sediments. Similar values are reported from an aquifer at a landfill site, where 98.3% of the cells were found attached (Röling et al. 2001). Also (Brad et al. 2008) emphasized the ecological role of sediments, which harbored cell numbers approximately one magnitude higher compared to groundwater.

Assuming a sediment surface area of 22 (coarse sand) to 450 cm^2 g^{-1} (fine sand) (Leichtfried 1985, Albrechtsen & Christensen 1994) and a mean area of 0.5 µm^2 covered by a typical bacterial cell in groundwater (Griebler et al. 2002), 4.4 x 10^9 to 9 x 10^{10} attached cells g^{-1} would be required to form a closed unicellular biofilm. In fact, the highest bacterial abundance determined from our sediment samples, *i.e.* at the lower plume fringe in Liner 1 (Fig. 2.3), accounts for a surface coverage of only 0.08 to 1.6%. This colonization density is comparable with other studies in aquifers (Griebler et al. 2001, Griebler et al. 2002) and marine sands (Weise & Rheinheimer 1977, Yamamoto & Lopez 1985). In such a case it is questionable, however, whether the term 'biofilm' is appropriate or whether we should rather refer to single cells and microcolonies. On the other hand, the maximal total direct cell counts of 4.1 to 4.8 x 10^7 cells g^{-1} wwt reported here for the upper sediment layers account for only ~20% of the maximally 2.4 x 10^8 bacterial 16S rRNA genes g^{-1} wwt detected via qPCR in similar depths upon sediment sampling in 2005 (Winderl et al. 2008). This number, consequently, would result in a surface coverage of 0.3 to 5.5%. It is likely that both quantification methods carry intrinsic biases, thus although they may be well suited to compare microbial distributions over depth, any absolute interpretations in terms of sediment carrying capacities should be treated with caution.

Indeed, the total cellular distribution did not reveal much about microbial activities and ongoing biodegradation processes. Hydrolysis rates of extracellular phosphatases (MUF-P substrate) and glucosidases (MUF-Glc substrate) (Freeman et al. 1995) revealed microbial activities 5 - 6 orders of magnitude higher in sediments compared to ground-water (Fig. 2.4). Again, highest values were obtained from the interface between the unsaturated and saturated zone, *i.e.* the capillary zone and upper plume fringe. As alkaline phosphatases are subject to catabolite repression by their product phosphate (Coolen & Overmann 2000), the high phosphatase activity in this zone corresponds well with the local decrease in phosphate concentration. Recycling of electron acceptors and mixing with electron donors from the contaminated area make the capillary fringe a favorable place for biodegradation (Sinke et al. 1998, Ronen et al. 2000). Higher activities of attached vs. suspended communities have been shown in a number of studies (Holm et al. 1992, Madsen & Ghiorse 1993, Albrechtsen & Christensen 1994, Lehman & O'Connell 2002). However, with β-glucosidase activity (Lehman & O'Connell 2002) also found the opposite pattern. Concluding, when comparing the gradients in figure 2.3 and 2.4 it can be seen that zones of enhanced microbial activity matched well zones with highest biomass concentration. High biomass zones in contaminated aquifers may thus be a good proxy for zones of pronounced NA activities.

2.4.5. Bacterial community shifts in sediment and water

Shifts in bacterial community composition over a depth transect of the Flingern BTEX-plume have been observed before for sediment samples taken in June 2005 (Winderl et al. 2008). Hence, an apparently highly specialized degrader population dominated by deltaproteobacterial and clostridal iron reducers, sulfate reducers and fermenters was found to dominate the lower plume fringe. In the present study, we repeated these analyses for selected sediment depths sampled in February 2006, but, more importantly, aimed to compare sedimentary community fingerprints to those obtained from corresponding groundwater samples. Depth-resolved fingerprinting of bacterial communities in water and sediments revealed that only ~50% of all detected T-RFs were shared between water and sediment samples. This is evidence that even if very close corresponding depths are sampled, the two compartments are partially separated on the microscale, and that not all aquifer microbes are equally detectable in both.

This comparison may provide valuable insights on the ecology of the different microbes, as exemplified e.g. by the comparative distribution pattern of the *Geobacter*-related T-RF (130 bp). This T-RF was practically not detected in groundwater, which may be attributed to the preference of these microbes to insoluble electron acceptors such as ferric iron and elemental sulfur, and hence to a lifestyle attached to mineral surfaces (Weber et al. 2006). In contrast, other microbes putatively more dependent on dissolved electron acceptors, such as the *Desulfocapsa*-relatives represented by the 159 bp T-RF, were consistently distributed with depth in both sediment and water samples, but always detected in higher ratios in groundwater. Moreover, the maximal abundance of this T-RF was surprisingly congruent to the peak of sulfide (Fig. 2.1B). This may be an indication that the distribution of these specific microbes is clearly correlated to the localization of sulfate reducing processes within the Flingern aquifer. We are fully aware that the inference of peak abundances in T-RFLP fingerprinting allow only for a semi-quantitative assessment of community composition and provide no absolute abundances of specific populations (Lueders & Friedrich 2003, Thies 2007, Hartmann & Widmer 2008). Nevertheless, we are confident to demonstrate important distinctions, but also similarities in depth-resolved microbial community distribution in sediments and water at the site using this approach. A significant correlation between community structure and degree of contamination was also observed in sediments of a leachate-impacted aquifer (Brad et al. 2008).

Over depth, the non-shared T-RFs appeared mainly in the deeper PAH zone (56% of all singletons). Thus, the upper zones dominated by the BTEX plume seem to impose much stronger selective pressures on the microorganisms, leading to a more uniform appearance of bacteria in both compartments. This was also reflected in the congruent depth-related bacterial community shifts identified in PCA, which supports both strategies to provide relevant information on depth-resolved microbial distribution patterns. Up to now, researchers have been well aware that only a minor fraction of total aquifer microbiota are found in groundwater itself (Alfreider et al. 1997, Lehman et al. 2001a, Griebler et al. 2002). Although we do not provide a direct quantitative comparison of microbial abundance in groundwater and sediments for the same sampling date, this is to our knowledge the first systematic evaluation of the congruence of physical-chemical and microbial community patterns found in both compartments.

We show that major microbial community indicators such as total diversity or the relative abundance of selected community members are distributed in similar, however not identical patterns. Sediment samples always seem to comprise most genetic information found also in the water, but this relation is not correct *vice versa*. Nevertheless, as ~50% of all T-RF peaks were shared between both compartments, both strategies can be considered appropriate to detect spatial and temporal distinctions in the microbial community composition at contaminated sites.

In summary, the data presented in this study emphasize the importance of considering both groundwater and sediment parameters for assessing natural attenuation potential and activities at organically contaminated aquifers. Unlike groundwater samples, which provide only a momentary snapshot of dissolved reactants and dependant microbes *in situ*, sediment analyses allowed more profound insights into the history of the aquifer and into long-term processes occurring at a site, not only in the aquatic, but also the sedimentary compartments.

2.5. References

Albrechtsen H-J, Christensen TH (1994) Evidence for microbial iron reduction in a landfill leachate-polluted aquifer (Vejen, Denmark). Applied and Environmental Microbiology 60:3920-3925

Albrechtsen H-J, Winding A (1992) Microbial biomass and activity in subsurface sediments from Vejen, Denmark. Microbial Ecology 23:303-317

Alfreider A, Krössbacher M, Psenner R (1997) Groundwater samples do not reflect bacterial densities and activity in surface systems. Water Research 31:882-840

Anderson RT, Lovley DR (1997) Ecology and Biochemistry of in Situ Groundwater Bioremediation. In: Gwynfryn Jones J (ed) Advances in Microbial Ecology, Vol 15. Plenum Press, New York, p 289-350

Anneser B, Einsiedl F, Meckenstock RU, Richters L, Wisotzky F, Griebler C (2008a) High-resolution monitoring of biochemical gradients in a tar oil-contaminated aquifer. Applied Geochemistry 23:1715-1730

Anneser B, Richters L, Griebler C (2008b) Application of high-resolution groundwater sampling in a tar oil-contaminated sandy aquifer. Studies on small-scale abiotic gradients. In: Candela L, Vadillo I, Elorza FJ (eds) Advances in Subsurface Pollution of Porous Media: Indicators, Processes and Modelling CRC press/Balkema, Taylor & Francis Group, London, UK, p 107-122

Appelo CAJ, Postma D (2005) Geochemistry, Groundwater and Pollution, 2 edn. Taylor & Francis Group, Leiden, The Netherlands

Bamforth SM, Singleton I (2005) Bioremediation of polycyclic aromatic hydrocarbons: current knowledge and future directions. Journal of Chemical Technology & Biotechnology 80:723-736

Bekins BA, Cozzarelli IM, Godsy EM, Warren E, Essaid HI, Tucillo M (2001) Progression of natural attenuation processes at a crude oil spill site: II. Controls on spatial distribution of microbial populations. Journal of Contaminant Hydrology 53:387-406

Brad T, van Breukelen BM, Braster M, van Straalen NM, Röling FM (2008) Spatial heterogeinity in sediment-associated bacterial and eukaryotic communities in a landfill leachate-contaminated aquifer. FEMS Microbiology Ecology 65:534-543

Brielmann H, Griebler C, Schmidt SI, Michel R, Lueders T (2009) Effects of thermal energy discharge on shallow groundwater ecosystems. FEMS Microbiology Ecology 68:273-286

Christensen TH, Bjerg PL, Banwart SA, Jakobsen R, Heron G, Albrechtsen H-J (2000) Characterizaion of redox conditions in groundwater contaminant plumes. Journal of Contaminant Hydrology 45:165-241

Christensen TH, Kjeldsen P, Bjerg PL, Jensen DL, Christensen JB, Baun A, Albrechtsen H-J, Heron G (2001) Biochemistry of landfill leachate plumes. Applied Geochemistry 16:659-718

Coolen MJL, Overmann J (2000) Functional Exoenzymes as Indicators of Metabolically Active Bacteria in 124,00-Year-Old Sapropel Layers of the Eastern Mediterranean Sea. Applied and Environmental Microbiology 66:2589-2598

Cozzarelli IM, Herman JS, Baedecker MJ, Fischer JM (1999) Geochemical heterogeneity of a gasoline-contaminated aquifer. Journal of Contaminant Hydrology 40:261-284

Eberhardt C, Grathwohl P (2002) Time scales of organic contaminant dissolution from complex source zones: coal tar pools vs. blobs. Journal of Contaminant Hydrology 59:45-66

Eckert P (2001) Untersuchungen zur Wirksamkeit und Stimulation natürlicher Abbauprozesse in einem mit gaswerkspezifischen Schadstoffen kontaminierten Grundwasserleiter. Ruhr-Universität Bochum, Bochum

Freeman C, Liska G, Ostle N, Jones S, Lock M (1995) The use of fluorogenic substrates for measuring enzyme activity in peatlands. Plant and Soil 175:147-152

Griebler C, Lueders T (2009) Microbial biodiversity in groundwater ecosystems. Freshwater Biology 54:649-677

Griebler C, Mindl B, Slezak D (2001) Combining DAPI and SYBR Green II for the Enumeration of Total Bacterial Numbers in Aquatic Sediments. International Review of Hydrobiology 86:453-465

Griebler C, Mindl B, Slezak D, Geiger-Kaiser M (2002) Distribution patterns of attached and suspended bacteria in pristine and contaminated shallow aquifers studied with an in situ sediment exposure microcosm. Aquatic Microbial Ecology 28:117-129

Hartmann M, Widmer F (2008) Reliability for detecting composition and changes of microbial communities by T-RFLP genetic profiling. FEMS Microbiology Ecology 63:249-260

Hazen T, Jiménez L, López de Victoria G, Fliermans C (1991) Comparison of bacteria from deep subsurface sediment and adjacent groundwater. Microbial Ecology 22:293-304

Hendel B, Marxsen J (1997) Measurement of Low-level Extracellular Enzyme Activity in Natural Waters Using Fluorigenic Model Substrates. Acta Hydrochimica et Hydrobiologica 25:253-258

Heron G, Crouzet C, Bourg ACM, Christensen TH (1994) Speciation of Fe(II) and Fe(III) in contaminated aquifer sediments using chemical extraction techniques. Environmental Science & Technology 28:1698-1705

Holm N, Cairns-Smith A, Daniel R, Ferris J, Hennet R, Shock E, Simoneit E, Yanagawa H (1992) Marine hydrothermal systems and the origin of life: future research. Origins of Life and Evolution of the Biosphere 22:181-242

Kao CM, Kota S, Ress B, Barlaz B, Borden RC (2001) Effects of subsurface heterogeneity on natural bioremediation at a gasoline spill site. Water Science and Technology 43:341-348

Kölbel-Boelke J, Hirsch P, Characklis W, Wilderer P (1989) Comparative physiology of biofilm and suspended microorganisms in the groundwater environment. In: Charcklis W, Wilderer P (eds) Structure and Fuction of Biofilms. John Wiley, Chichester

Kostka J, Roychoudhury A, Van Chappellen P (2002) Rates and controls of anaerobic microbial respiration across spatial and temporal gradients in saltmarsh sediments. Biogeochemistry 60:49-76

Lehman RM, Colwell FS, Bala GA (2001a) Attached and Unattached Microbial Communities in a Simulated Basalt Aquifer under Fracture- and Porous-Flow Conditions. Applied and Environmental Microbiology 67:2799-2809

Lehman RM, O'Connell SP (2002) Comparison of Extracellular Enzyme Acitivities and Community Composition of Attached and Free-Living Bacteria in Porous Medium Columns. Applied and Environmental Microbiology 68:1569-1575

Lehman RM, Roberto FF, Earley D, Bruhn DF, Brink SE, O'Connell SP, Delwiche ME, Colwell FS (2001b) Attached and unattached bacterial communities in a 120-meter corehole in an acidic, crystalline rock aquifer. Applied and Environmental Microbiology 67:2095-2106

Leichtfried M (1985) Organic Matter in Gravel Streams (Project RITRODAT-LUNZ). Verhandlung Internationale Verunreinigung Limnologie 22:2058-2062

Lindahl V, Bakken LR (1995) Evaluation of methods for extraction of bacteria from soil. FEMS Microbiology Ecology 16:135-142

Lovley DR (2001) Anaerobes to the rescue. Science 293:1444-1446

Lovley DR, Chapelle FH (1995) Deep Subsurface Microbial Processes. Reviews of Geophysics 33:365-381

Lovley DR, Phillips EJP (1987) Rapid Assay for Microbially Reducible Ferric Iron in Aquatic Sediments. Applied and Environmental Microbiology 53:1536-1540

Lovley DR, Phillips EJP (1988) Novel mode of microbial energy metabolism: organic carbon oxidation coupled to dissimilatory reduction of iron or manganese. Applied and Environmental Microbiology 54:1472-1480

Lueders T, Friedrich MW (2002) Effects of Amendment with Ferrhydrite and Gypsum on the Structure and Activity of Methanogenic Populations in Rice Field Soil. Applied and Environmental Microbiology 68:2484-2494

Lueders T, Friedrich MW (2003) Evaluation of PCR Amplification Bias by Terminal Restriction Fragment Length Polymorphism Analysis of Small-Subunit rRNA and mcrA Genes by Using Defined Template Mixtures of Methanogenic Pure Cultures and Soil DNA Extracts. Applied and Environmental Microbiology 69:320-326

Madsen EL, Ghiorse WC (1993) Groundwater microbiology: subsurface ecosystem processes. In: Aquatic Microbiology - An Ecological Approach. Blackwell Scientific publications, p 167-213

McGuire JT, Smith EW, Long DT, Hyndman DW, Haack SK, Klug MJ, Velbel MA (2000) Temporal variations in parameters reflecting terminal-electron-accepting processes in an aquifer contaminated with waste fuel and chlorinated solvents. Chemical Geology 169:471-485

Nebe-von-Caron G, Stephens PJ, Hewitt CJ, Powell CR, Badley RA (2000) Analysis of bacterial function by multi-colour fluorescence flow cytometry and singel cell sorting. Journal of Microbiological Methods 42:97-114

Paissé S, Coulon F, Goñi-Urriza M, Peperzak L, McGenity TJ, Duran R (2008) Structure of bacterial communities along a hydrocarbon contamination gradient in a coastal sediment. FEMS Microbiology Ecology 66:295-305

Roden EE, Zachara JM (1996) Microbial Reduction of Crystalline Iron(III) Oxides: Influence of Oxide Surface Area and Potential for Cell Growth. Environmental Science & Technology 30:1618-1628

Röling WFM, Van Breukelen BM, Braster M, Lin B, Van Verseveld HW (2001) Relationships between Microbial Community Structure and Hydrochemistry in a Landfill Leachate-Polluted Aquifer. Applied and Environmental Microbiology 67:4619-4629

Ronen D, Scher H, Blunt M (2000) Field observations of a capillary fringe before and after a rainy season. Journal of Contaminant Hydrology 44:103-118

Sinke AJC, Dury O, Zobrist J (1998) Effects of a fluctuating water table: column study on redox dynamics and fate of some organic pollutants. Journal of Contaminant Hydrology 33:231-246

Stumm W, Morgan JJ (1996) Aquatic Chemistry. John Wiley & Sons, New York

Szabo G, Prosser SL, Bulman RA (1990) Determination of the adsorption co-efficient Koc of some aromatics for soils by RP-HPLC on two immobilized humic acid phases. Chemosphere 21:777-788

Thies JE (2007) Soil Microbial Community Analysis using Terminal Restriction Fragment Length Polymorphisms. Soil Science Society of America Journal 71:579-591

Tuccillo ME, Cozzarelli IM, Herman JS (1999) Iron reduction in the sediments of a hydrocarbon-contaminated aquifer. Applied Geochemistry 14:655-667

Ulrich GA, Breit GN, Cozzarelli IM, Suflita JM (2003) Sources of Sulfate Supporting Anaerobic Metabolism in a Contaminated Aquifer. Environmental Science & Technology 37:1093-1099

Ulrich GA, Krumholz LR, Suflita JM (1997) A rapid and simple method for estimating sulfate reduction activity and quantifying inorganic sulfides. Applied and Environmental Microbiology 63:4626

Ulrich GA, Martino D, Burger K, Routh J, Grossman EL, Ammerman JW, Suflita JM (1998) Sulfur cycling in the terrestrial subsurface: Commensal interactions, spatial scales, and microbial heterogeneity. Microbial Ecology 36:141-151

USACE (2002) Soil vapor extraction and bioventing Engineering and Design. US Army Corps of Engineers

Weber KA, Achenbach LA, Coates JD (2006) Microorganisms pumping iron: anaerobic microbial iron oxidation and reduction. Nature Reviews Microbiology 4:752-764

Weise W, Rheinheimer G (1977) Scanning electron microscopy and epifluorescence investigation of bacterial colonization of marine sand sediments. Microbial Ecology 4:175-188

Widdel F, Bak F (1992) Gram negative mesophilic sulfate reducing bacteria. In: Balows A, Trüper HG, Dworkin M, Harder W and Schleifer K-H (eds). The Prokaryotes, Springer: New York, NY, pp 3352–3378.

Winderl C, Anneser B, Griebler C, Meckenstock RU, Lueders T (2008) Depth-resolved quantification of anaerobic toluene degraders and aquifer microbial community patterns in distinct redox zones of a tar oil contaminant plume. Applied and Environmental Microbiology 74:792-801

Yamamoto N, Lopez G (1985) Bacterial abundance in relation to surface area and organic content of marine sediments. Journal of Experimental Marine Biology and Ecology 90:209-220

Zhou J, Bruns MA, Tiedje JM (1996) DNA recovery from soils of diverse composition. Applied and Environmental Microbiology 62:316-322

3. Collapse and recovery of intrinsic toluene degradation – transient hydraulic conditions control natural attenuation in a tar oil contaminated sandy aquifer

Groundwater systems are considered stable in their environmental conditions. However, frequently they are subject to pronounced hydrological dynamics. Little is know to what extent hydrodynamics influence aquifers microbiology. Natural attenuation (NA) of an organic contaminant plume in the porous sandy aquifer of a former gasification plant was assessed for the period of four years. Vertical gradients of selected physical-chemical and microbial variables, including the concentrations of major contaminants (BTEX, naphthalene) and electron acceptors, reaction products, and total bacterial numbers have been recorded at an unprecedented spatial resolution in the centimetre scale. Significant changes of the small-scale biogeochemical gradients were observed closely linked to fluctuations of the groundwater surface. Biodegradation hot spots, always located in the plume fringe zones, followed the vertical dynamics of the plume, hinting at a broad distribution of bacterial degrader populations tolerant to changing redox conditions and actively contributing to contaminant transformation the moment conditions turn favourable. However, stable isotope analysis revealed conclusive evidence for an interim breakdown of *in situ* biodegradation. These observations of a frequent disturbance or even disruption of efficient natural contaminant removal by highly specialized anaerobic degrader populations clearly asks for an improved understanding of degrader ecology in the light of NA becoming the most popular remediation concept.

3.1. Introduction

During industrialization big cities have generally built coal gasification plants to support municipal gas supply. By different production lines these plants have produced diverse compound spectra consisting of e.g. BTEX (benzene, toluene, ethylbenzene, xylene), PAHs (polycyclic aromatic hydrocarbons), heterocyclic compounds, and tar oil which is a mixture of all (Meckenstock et al. 2010). For a long time it was a common practice to dump such waste directly into ponds or soil. In the 1950s, many plants were closed after having been operated for more than 100 years. Spilled aromatic hydrocarbons lead to heavy contamination of soil and groundwater at these sites. Owing for their low density, BTEX compounds and low molecular PAHs are lighter than water and, hence, floating as light non-aqueous phase liquids (LNAPLs) on top of the groundwater table. In response to fluctuations of the groundwater table and groundwater flow, such LNAPLs transversely spread within the aquifer. In sandy aquifers, the contaminations usually form vertically thin plumes of no more than 1 - 2 m thickness close to the groundwater table. In case, compounds are heavier than water, e.g. high molecular PAHs and tar oil (dense non-aqueous phase liquids DNAPLs), these penetrate the aquifer down to deeper zones, producing very thick contamination plumes.

The load of organic material leads generally to a depletion of dissolved oxygen; these anoxic conditions prevail in aquifers at former coal gasification sites. Aerobic degradation processes in contaminated areas may contribute only in the aerated unsaturated zone and at the capillary fringe. In the subjacent areas sulfate and ferric iron are the most important electron acceptors (Lewandowski & Mortimer 2004, Anneser et al. 2010). Natural aquifer microbial communities have been shown to carry a great potential for anaerobic degradation of aromatic hydrocarbons (Meckenstock et al. 2004, Hendrickx et al. 2005). However, the environmental factors controlling efficient NA processes *in situ* are far from being well understood.

In lab experiments and reactive transport modeling, transverse dispersion was proven to be a major factor controlling biodegradation in porous sediments (Prommer et al. 2002, Bauer et al. 2008). Consequently, main biodegradation activities are located at the plume fringes, where mixing of electron acceptors and electron donors provides essential conditions for degraders (Mayer et al. 2001, Vencelides et al. 2007). Similar to lake and marine sediments, steep geochemical and microbial gradients have been demonstrated for the fringe zones of hydrocarbon plumes (van Breukelen et al. 2004, Bauer et al. 2009, Anneser et al. 2010). Another factor posing significant influence on contaminant biodegradation was shown to be sediment heterogeneity (Cozzarelli et al. 1999, Bauer et al. 2009). Zones of increased hydraulic conductivity generate, via flow focusing, increased flow velocity and improved mixing, areas of effective biodegradation.

Hardly considered so far was the possible influence of transient hydraulic conditions to biodegradation. Porous aquifers are from an ecological perspective generally considered environmentally static systems. However, there is a number of reports emphasizing hydrodynamics in aquifers caused by (i) pronounced recharge events and changes in the groundwater table (McGuire et al. 2005, Wilhartitz et al. 2009, Yagi et al. 2010), (ii) tides (Robinson et al. 2009), (iii) changes in the direction of the groundwater flow (Rein et al. 2009) or (iv) density-dependent changes in the saturated zone. It may be hypothesized that hydraulic dynamics have a positive effect on biodegradation by improving the distribution of the contaminants to less contaminant zones. At the same time, the frequent shift or destruction of physical-chemical gradients established may pose a serious disturbance to the mainly attached and highly specialized degrader populations. Owing to fluctuations, they may repeatedly be exposed to either groundwater carrying electron acceptors but depleted in contaminants or, *vice versa*, contain a high and possibly toxic load of organics but no dissolved electron acceptors.

In this study, we addressed the positive vs. negative effects of transient hydraulic conditions on the overall biodegradation of toluene in a porous aquifer at a former gasification plant. Vertical gradients of dominating contaminants, electron acceptors, metabolic products, and bacterial counts have been followed over a period of four years at an extraordinary high resolution (3 - 10 cm). Compound-specific isotope analysis allowed identifying and specially resolving toluene degradation as well as sulfate reduction. The collapse and recovery of toluene degradation occasionally captured by our series of sampling events at weekly to monthly intervals allows discussion on the resistance and resilience of degrader communities with respect to hydrodynamics.

3.2. Materials and methods

3.2.1. Study site and groundwater sampling

The study site is a quaternary, homogenous sandy aquifer at a former gasification plant in Düsseldorf-Flingern, Germany. On a local scale, groundwater is heavily polluted with aromatic hydrocarbons from a tar oil source dominated by BTEX and naphthalene among others (e.g. acenaphthene, fluorene, dibenzofurane). More details on the study site are provided by (Anneser et al. 2008a, Anneser et al. 2008b). Sampling of groundwater was conducted over a period of four years to compare long- and short-term spatial and temporal dynamics of NA. Groundwater was collected from a high-resolution multi-level well (HR-MLW) with a vertical resolution of 3 to 30 cm via simultaneously pumping of up to 32 sampling ports at low rates (1.5 mL min^{-1}), thus minimizing the risk of mixing water from different depths. For further details on sampling procedures see (Anneser et al. 2008a, Anneser et al. 2008b).

The groundwater table was daily measured via data logger in close vicinity of the sampling well (Fig. 3.1).

3.2.2. Physical-chemical characteristics of groundwater

Groundwater was collected in 100 mL pre-washed narrow neck-glass bottles allowing overflow and immediately processed or preserved on site for a number of biotic and abiotic parameters. Conductivity, pH, redox potential, iron, sulfide, major ions, and aromatic hydrocarbons were determined as described elsewhere (Anneser et al. 2008a, Anneser et al. 2008b, Anneser et al. 2010).

3.2.3. Quantification of cell numbers and bacterial carbon production (BCP)

Subsamples of groundwater dedicated to total bacterial cell counts were fixed with glutardialdehyde (2.5% final concentration) and stored at +4°C prior to quantification using flow cytometry (Berney et al. 2008, Hammes et al. 2008).

BCP was estimated via incorporation of [^{3}H]-leucine by adapting the protocols of Kirchman (1993), Kisand & Noges (2004) and Brielmann et al. (2009). Actual incorporation of label into proteins was estimated to account for 10% of the total bacterial [^{3}H]-leucine uptake. According to this assumption, BCP was calculated as described in Kirchman & Ducklow (1993).

3.2.4. Stable isotope analysis

In order to determine the isotopic composition of sulfate in groundwater, subsamples were fixed in a 20% (w/v) Zn-acetate solution for precipitation of sulfide. Stable sulfur ($^{34}S/^{32}S$) and oxygen ($^{18}O/^{16}O$) isotopes in the remaining sulfate were determined after filtration according to Anneser et al. (2008a).

Toluene $^{13}C/^{12}C$ isotope ratios were analyzed on NaOH-preserved groundwater samples with gas chromatography-combustion-isotope ratio mass spectrometry (GC-C-IRMS, Thermo Fisher, Germany) via purge & trap (Kuhn et al. 2009), with the gas chromatographic separation adapted to BTEX. A DB624-column (Supelco, USA) and an optimized temperature program (3.07 min at 60°C, 8.1°C min^{-1} to 129°C, 1°C min^{-1} to 134°C, 20°C min^{-1} to 180°C, 50°C min^{-1} to 230°C, hold 5 min) were used.

The deviations for duplicate toluene as well as sulfate stable isotope measurements were much lower than the total instrumental uncertainty, and therefore the analytical standard deviation for both methods was set to 0.5‰ (Sherwood Lollar et al. 2007).

3.2.5. Dynamic expansion of the BTEX plume

Information about the BTEX concentrations gained from conventional multi-level wells (C-MLWs) at the site was used for illustrating the dynamic expansion of the BTEX plume for three different sampling occasions in the year 2007, 2008 and 2009. BTEX concentrations at sampling points were spatially interpolated creating a contour plot using the software Surfer 9 with the Kriging approach (Schäfer et al. 2004).

3.3. Results and discussion

Porous aquifers are generally considered systems of constant environmental conditions, which make them the preferred sources for continuous extraction of high quality water, e.g. for drinking water production. The picture of the stable environment mainly derives, in comparison with surface aquatic systems, from the moderate temperature fluctuations and low organic carbon as well as microbial biomass concentrations observed. Without doubt, these systems may face considerable dynamics in hydrological patterns, such as groundwater table changes, which, however, have rarely been evaluated for its potential control of biological features (Vroblesky & Chapelle 1994, Cozzarelli et al. 1999, McGuire et al. 2000, McGuire et al. 2002, McGuire et al. 2005). Especially where considering NA as sole remediation strategy, the influence of hydrodynamics on contaminant biodegradation is of great importance. We therefore selected a sandy aquifer at a former coal gasification plant with a long history of hydrocarbon waste disposal. For capturing long-term and short-term temporal and spatial dynamics in various physical-chemical and microbial parameters in total ten sampling campaigns were conducted between September 2005 and June 2009. The data depicted are exclusively derived from sampling groundwater out of a specially designed high-resolution multi-level monitoring well (Anneser et al. 2008a, Anneser et al. 2008b). Additional information obtained from infrequent sediment sampling campaigns are presented in detail elsewhere (Anneser et al. 2010, Meckenstock et al. 2010) and considered for discussion of groundwater patterns.

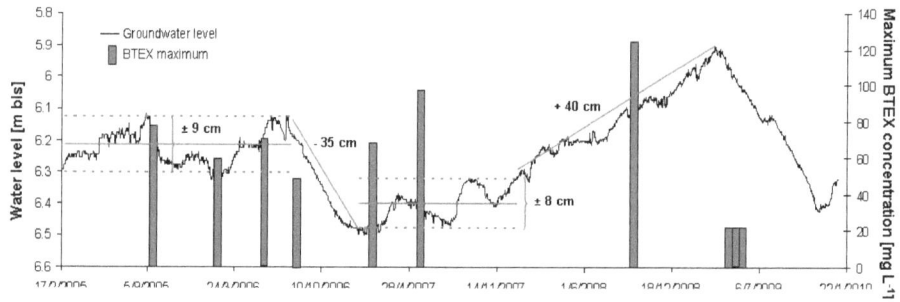

Figure 3.1: Dynamics of the groundwater table from continuous recording at a fully screened well in the direct vicinity of the high-resolution multi-level well. Important variations are highlighted. Grey bars indicate dates of groundwater sampling together with the maximum BTEX concentration determined along the vertical profiles.

3.3.1. Chemical and microbiological indicators of biodegradation

Presence and allocation of zones of high bioactivity could be elucidated and followed monitoring steep and small scale vertical gradients of contaminants (electron donors) and electron acceptors, reaction products and further redox-sensitive parameters. Concentration profiles of individual aromatic hydrocarbons are delineating a distinct contaminant plume, which contains toluene as main constituent, as well as xylenes, ethylbenzene, naphthalene and benzene, exhibiting a restricted vertically down-gradient expansion of less than 1 m starting at the groundwater table (Fig. 3.2).

The high concentrations of dissolved sulfide located in the zones of opponent steep sulfate and BTEX gradients corroborate our assumption that biodegradation of BTEX compounds and the physical-chemical gradients detected in the groundwater are mainly attributed to microbial sulfate reduction as the most prominent redox process. This indirect indication is substantiated by compound-specific isotope analysis (CSIA) of sulfate and toluene.

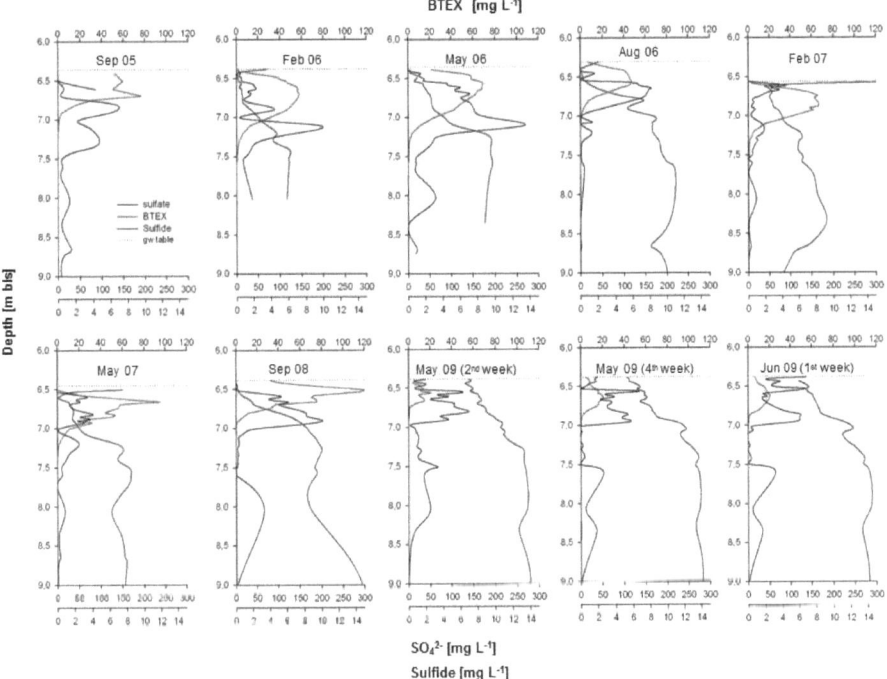

Figure 3.2: Vertical distribution of BTEX, sulfate and sulfide from all sampling campaigns over the observation period of four years. Values are means of duplicate measurements. Data from February and August 2006 have already been published elsewhere (Anneser et al. 2010) and (Anneser et al. 2008a), respectively. Short-term sampling was performed in 2009 at a time range of one and two weeks, respectively.

Stable sulfur and oxygen isotopes of sulfate reinforce microbial sulfate reduction activity taking place mainly in the plume fringe zones, visible from the steep gradients of isotopic enrichment towards the plume center (Fig. 3.3A). Plotting $\delta^{34}S$ versus $\delta^{18}O$ values indicates a common isotopically similar sulfate source, with the exception of individual samples collected at the capillary fringe (Fig. 3.3B). Here, the distinct sulfur and oxygen isotope patterns point at repeated recycling of the electron acceptor sulfate via sulfide oxidation (Fig. 3.3B) (Anneser et al. 2010). The biological transformation of contaminants could be clearly demonstrated by CSIA of toluene, which proved biodegradation to be most pronounced at the plume fringes, as is depicted from the years 2006, 2007 and 2009 (Fig. 3.3C). Now, the gradients show the typical isotopic enrichment from the plume center towards the non contaminated surrounding. Stable isotope gradients of toluene missing in year 2008 are related to a collapse of toluene degradation (for more details see below).

Besides sulfate reduction, indication was collected for microbial iron reduction to take place all across the investigated aquifer section, with particular evidence for the upper and lower plume fringes, presented by gradients of dissolved Fe(II) (Anneser et al. 2008a, Anneser et al. 2008b, Anneser et al. 2010, Meckenstock et al. 2010). In terms of biodegradation, iron reduction seems to play an inferior role. However, this point awaits further evaluation in the future (Anneser et al. 2008a, Anneser et al. 2010). Over the entire period of investigations, reducing conditions prevailed in the saturated aquifer right below the capillary fringe. Nitrate occurred only occasionally at concentrations less than 2 mg L^{-1} (Anneser et al. 2008a, Anneser et al. 2008b). Moreover, we did not obtain indication for manganese reduction and methanogenesis (Anneser et al. 2008a).

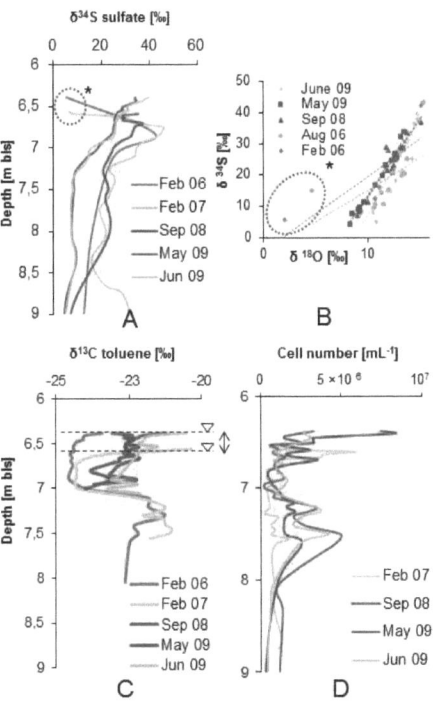

Figure 3.3: Vertical distribution of the $\delta^{34}S$-values of sulfate (A), the relationship between $\delta^{34}S$ and $\delta^{18}O$ in sulfate (B), the $\delta^{13}C$-values of toluene (C), and total bacterial cell counts (D) in groundwater. Values are means of duplicate (sulfate) and triplicate (toluene, bacterial numbers) measurements. The asterisk denotes $\delta^{34}S$-values indicating a different source of sulfate including its origin from sulfide oxidation. BTEX, sulfate and sulfide data from February and August 2006 have already been published elsewhere (Anneser et al. 2010) and (Anneser et al. 2008a), respectively.

Directly at the capillary fringe which constitutes the upper BTEX plume fringe, the highest bacterial abundance could be found (Fig. 3.3D). With an increasing degree of contamination towards the plume centre, a decrease in biomass was repeatedly observed at all sampling occasions, while the transition to less contaminated groundwater at the lower plume fringe returns more favorable conditions, as indicated by a moderate but distinct increase in total cell counts. In deeper zones of the aquifer (~7.5 m below land surface (bls)) the bacterial abundance showed again a peak. Here, a second plume of mainly naphthalene, acenaphthene and fluorene was recognized predominantly with sediment analysis (Anneser et al. 2010) and occasionally also from groundwater analysis. In this deeper zone of the aquifer, maximum contaminant concentrations in groundwater ranged between 1.1 mg L^{-1} in 2006 and 2.5 mg L^{-1} in 2008 (data not shown).

3.3.2. Long-term dynamics of biochemical gradients

We propose that hydrological dynamics at the Düsseldorf-Flingern aquifer are, in interplay with the microbial communities, the major driving force for the distribution of contaminants, electron acceptors, reaction products, and, consequently, the dynamics found with microbial biodegradation focusing mainly on toluene transformation.

The hydrodynamics present are assumed to be driven by local and regional precipitation events, the connectivity to the river Rhine system, and sediment heterogeneities between the contaminant source and the site of observation, *i.e.* the high resolution multi-level well (estimated distance of 15 - 25 m; Anneser et al. 2008a). Periods of strong hydrodynamics were delineated from pronounced groundwater table changes. However, as most of the local surface area is sealed and the unsaturated zones extents more than 6 m in depth, an immediate impact of local recharge is unlikely, and may take effect with a time lag. Hence, the oscillations of the groundwater table is a clear sign for pronounced hydrodynamics but could not directly be correlated to local precipitation events and water levels of the associated river Rhine system. Similarly, the fact that changes of the groundwater surface have influenced dissolution processes at the source, may have impacted measurements at our sampling site with a delay of 7 - 50 days, accounting for a groundwater flow velocity of 0.1 - 2 m day^{-1} (Anneser et al. 2008a).

During the initial phase of our study, March 2005 and September 2006, the moderate fluctuations of the groundwater table, *i.e.* ±9 cm (Fig. 3.1), already went along with substantial changes in the vertical distribution of the BTEX compounds (Fig. 3.2). Major peaks of sulfide were exclusively found associated with the lower plume fringe. From August 2006 on, the hydraulic head started to drop continuously until the end of the year by 35 cm from 6.14 to 6.49 m bls (Fig. 3.1). The plume followed vertically down-gradient staying constant in its expansion (Fig. 3.2). During this time highest sulfide concentrations were suddenly found at the upper plume fringe matching the former position of the lower fringe. Now, the original prominent lower fringe sulfide peak was missing. In the following months, only moderate groundwater table changes occurred, *i.e.* ±8 cm. In May 2007 sulfide peaks were present in the upper and lower plume fringe zone (Fig. 3.2), and in 2008 the maximum sulfide concentration again was found with the lower part of the BTEX plume. From January 2008 on, the groundwater table constantly rose until March 2009, leading to a concomitant upward movement of the plume, as indicated by the September 2008 data (Fig. 3.2), exhibiting shortly a hydraulic head 40 cm above that recorded during the sampling event in February 2007. As a general trend, the maximum BTEX concentrations detected in the plume center increased with a rising groundwater table and, *vice versa* showed a decreasing trend when the groundwater table declined. The zone of high dissolved sulfide concentrations seemed to be able to follow the moving lower plume fringe to a certain extent, but sulfide concentrations dropped with changing environmental conditions. The spatially distinct distribution of sulfate reducing bacterial contaminant degrader populations attached to the sediments may explain these dynamic patterns best. Moderate fluctuations of the groundwater table keep the lower plume fringe in a defined zone densely colonized by highly specialized degrader populations. Indeed, sediment bacterial community analysis revealed the quantitative importance of *Desulfocapsa* spp. and clostridial sulfate reducers in the lower plume fringe zone in February 2006 (Anneser et al. 2010). Moreover, in the lower plume fringe area, almost every second cell was found carrying the benzylsuccinate synthase *(bssA)* gene responsible for anaerobic toluene and xylene degradation (Winderl et al. 2008). Although the dominant degraders of the lower fringe were also present at the upper plume fringe, their relative abundance was comparably low, being accompanied by other quantitatively more important groups of iron and sulfide oxidizers (Winderl et al. 2008, Pilloni et al. 2011). The very high BTEX concentrations observed in September 2008 followed by the extraordinary low concentration only a few months later in 2009 (Fig. 3.2) attracted specific attention.

3.3.3. Collapse of toluene degradation

Evaluating the patterns of the individual BTEX compounds, it became obvious that only toluene and xylene dramatically increased in concentration in 2008, while other compounds remained unchanged. This became obvious looking at the ratios of toluene to other compounds of interest (Fig. 3.4).

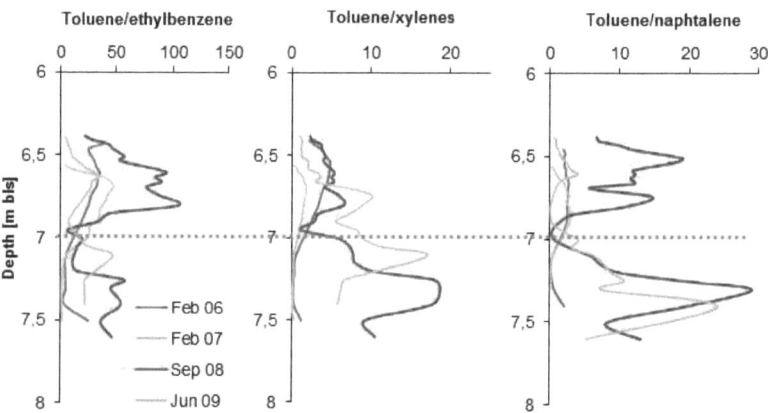

Figure 3.4: Vertical distribution of ratios of toluene to ethylbenzene, to xylenes, and to naphthalene, respectively. Note, contaminant concentrations at greater depths (below dotted line) fell close to detection limit. Consequently small differences in concentrations lead to huge differences in ratios and these data need to be interpreted with caution.

The maximum toluene concentration differed remarkably with 97 mg L^{-1} in 2008 and only 18.3 mg L^{-1} in June 2009. The least changes were found with benzene, ethylbenzene and naphthalene. In detail, the maximum concentrations of benzene and naphthalene, with 0.08 - 0.14 mg L^{-1} and 5.3 - 7.5 mg L^{-1}, respectively, did not follow the huge shift in toluene concentration. Changes in ratios of toluene to these compounds found at the lower plume fringe are mainly a result of very low concentrations. Here, minor changes in concentrations lead to pronounced changes in ratios. As could be supported by data from several conventional multi-level wells (C-MLWs), changes in BTEX concentrations along the vertical profile could be related to changes in plume length (Fig. 3.5).

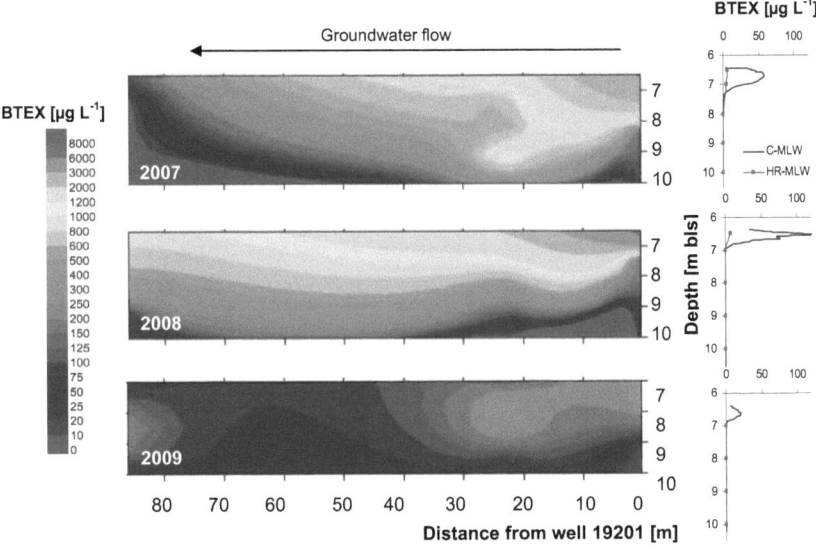

Figure 3.5: Two dimensional distribution of BTEX compounds. Vertically resolved data from three conventional multi-level wells, sampled in February 2007, August 2008 and August 2009, are spatially extrapolated by a Surfer plot (kriging approach). Selected vertical profiles depicted to the right, directly compare contaminant concentration data of individual depths from the conventional multi-level well (well C-MLW 19201) and the high-resolution multi-level well (HR-MLW), respectively. C-MLW data were provided by the Stadtwerke Düsseldorf.

With increasing contaminant concentrations in the plume, the sulfate availability declined, sometimes to nearly full sulfate depletion in the plume center and at the upper plume fringe as observed in 2006 and 2008 (Fig. 3.2). The high amounts of sulfide and concomitant low sulfate concentrations measured in September 2008 pointed at an unaltered high sulfate reducing activity. However, this time the toluene concentration was the highest ever measured since 2005. Sulfate isotope analysis also underlined sulfate reduction to occur (Fig. 3.3A), but the sulfate from plume core samples was found less isotopically enriched, when compared to the years before. In 2009 the data differ from the previous years and, hence, denote a lower sulfate reduction peaking at the top most sample in the upper plume fringe (Fig. 3.3A). The most striking evidence for an intermediate collapse of toluene degradation was obtained from stable carbon isotope profiles of toluene. The distinct $^{13}C/^{12}C$ isotope pattern of toluene at the plume fringes as observed in 2006 and 2007 could not be found in 2008. At that time stable isotope values, although overall enriched by about 2‰, did not indicate spatially distinct toluene degradation. In May 2009, a gradient in toluene isotope signatures established again at the upper plume fringe (Fig. 3.3C).

Independent evidence for the collapse of toluene, and possibly also xylenes degradation in 2008, originates from the analyses of sediment bacterial communities, as well as the quantitative evaluation of the *bssA* genes. Pyrosequencing of 16S rRNA genes revealed that along with the high toluene concentrations in 2008, an initial deltaproteobacterial degrader population belonging to the *Desulfobulbaceae* dramatically decreased in relative abundance and was quantitatively replaced by distinct clostridial degraders (*Desulfosporosinus* spp. of the *Peptococcaceae*) (Pilloni 2011). Analysis of *bssA* genes revealed a tremendous reduction in gene copies in 2008, which recovered to some degree in 2009 (Pilloni 2011). Interestingly, a decline in the total bacterial biomass in 2008, indicated by 16S rRNA gene copy numbers (Pilloni et al. 2011), could not be supported by total numbers of bacterial cells as assessed via flow cytometry (Fig. 3.3D). Dynamics inside bacterial communities without changes of the overall bacterial biomass, most probably determined by the system's carrying capacity, are not uncommon (Elliot et al. 2010, Yagi et al. 2010).

3.3.4. Short-term plume dynamics

Within a time span of less than three weeks in May and June 2009 the contaminant plume was evaluated three times for its short-term dynamics in physical-chemical and microbial patterns. Maximum BTEX concentrations during this period varied only slightly from 21.3 mg L^{-1} in May to 20.3 mg L^{-1} in June 2009 (Fig. 3.2). Only moderate changes occurred in plume thickness, as well as the vertical distribution of sulfate and sulfide. The latter showed increasing concentrations at the uppermost sampling point close to the groundwater table, indicating sulfate reduction going on in the upper plume fringe (Fig. 3.2). This picture is supported by the sulfate isotope data (Fig. 3.3A), which revealed a continuous enrichment towards the groundwater surface. At that time, no evidence from isotope data for a pronounced import of sulfate from the unsaturated zone or the sulfide oxidation in the capillary fringe was found. Sulfate, at that period in 2009, was available in sufficient amounts all over the plume cross section (Fig. 3.2). The vertical distribution of dissolved Fe(II), originating from Fe(III)-reduction, showed no pronounced dynamics within the three weeks (data not shown). Isotopic evidence for biodegradation was obtained with respect to toluene (Fig. 3.3C) and *m/p*-xylene (data not shown), with most pronounced enrichment gradients at the upper plume fringe. A clear change was observed at the lower plume fringe. Here, the gradients indicate no degradation in May, but re-established activities in June (Fig. 3.3C).

No differences were found with the total cell counts in groundwater during this period (Fig. 3.3D).

3.3.5. Transient hydraulic conditions and its potential influence on NA

Mixing via transverse and longitudinal dispersion is known to be one of the driving factors for biodegradation in porous aquifers (Thullner et al. 2002, Van Breukelen & Griffion 2004, Tuxen et al. 2006, Rees et al. 2007, Bauer et al. 2008). As already mentioned in the introduction section, transient hydraulic conditions in terms of groundwater table fluctuations may be suggested to be both, stimulating and inhibiting for the overall biodegradation. On one hand stimulating, because it may lead to the distribution of contaminants over a larger volume in the aquifer. This clearly improves the ratio of available electron acceptors and the contaminants. Moreover, it may contribute to lower contaminant concentrations, less toxic to individual members of the microbial community. In detail, along with groundwater table fluctuations, contaminants may be spread into zones with oxidative potential or are partly consigned into the capillary fringe and as such remain sorbed in the unsaturated zone after return of the groundwater table to a lower level and/or partition into soil atmosphere. On the other hand, we may consider that highly specialized assemblages of degrader populations required for the transformation of many organic contaminants, are sensitive to pronounced changes of environmental conditions such as redox states. The major microbial biomass ($\geq 95\%$), as known, is found attached to sediment grains (Alfreider et al. 1997, Griebler 2001, Anneser et al. 2010) and does, consequently, not constitute a mobile and flexible unit. Considering additionally the relatively slow doubling times of groundwater bacteria, that range from months to years (Mailloux & Fuller 2003, Griebler & Lueders 2008), they can not react as fast as the hydraulics change. A pronounced transversal or longitudinal shift of the plume due to transient hydraulic conditions may put the degraders under the periodic pressure of changing electron donor and acceptor concentrations, changing redox conditions, and possibly temporal gradients of toxicity. Taking into account, the comparably slow establishment or re-establishment of anaerobic bacterial degrader populations in aquifers, short-term hydrodynamic disturbances may have a long lasting effect to biodegradation. The biology is always lacking behind and constitutes beside the dispersion (mixing) of electron donor(s) and acceptor(s) another important limiting factors for biodegradation processes in porous aquifer.

A general conclusion if hydrodynamics in porous aquifers enhance or rather impair NA is not possible. At the Düsseldorf-Flingern site, characterized by its sulfate reducing degrader populations, evidence for a stimulating effect on biodegradation comes from isotope data which hint at frequent recycling of the electron acceptors sulfate and ferric iron (Fig. 3.3B; Anneser et al. 2008a, Anneser et al. 2010). However, the experience gained over the period of almost five years also points at pronounced inhibiting effects, culminating in the collapse of biodegradation of the major contaminant toluene.

Although the entire picture of NA at the Düsseldorf-Flingern aquifer is far from complete, there are a few facts we propose. (i) Highly specialized, sulfate reducing degrader populations are distributed all over the permanently and infrequently contaminated areas, with their highest relative abundance in the transition zones, *i.e.* the plume fringes. (ii) These populations are able to withstand interim variations in environmental conditions. In times of stability, they increase in relative abundance in zones of favorable living conditions building up a constant biomass confined by the carrying capacity of the system. (iii) At times of unfavorable conditions, e.g. interim shift of the plume fringes zones to other positions, they are suggested to persist in a low active or inactive stage for the time span of several months and more. With the return of favorable conditions, they continue their contribution to NA of the contaminants. (iv) However, our case study, for the first time, provides striking evidence that a hydrodynamic disturbance, such as a groundwater table rise of several decimeters with its physical-chemical consequences has the potential to directly or indirectly lead to the collapse of biodegradation of individual compounds (September 2008) and with a delay of several months to the replacement of the dominant degrader population (*Desulfobulbaceae*) by a new one (*Peptococcaceae*) (Pilloni 2011).

At the moment, it cannot be answered if first the abundant degrader population of the *Desulfobulbaceae* (Pilloni 2011) collapsed which lead to the extraordinary high toluene concentrations accompanied by the indistinct toluene carbon isotope signatures or, the other way round, the extraordinary high toluene concentrations and changes in redox conditions caused by transient hydraulic conditions poisoned the sulfate reducing *Desulfobulbaceae* population clearing the way for the establishment of a new, functionally redundant, degrader population within the *Peptococcaceae*. Factors such as (i) contaminant toxicity, (ii) sensitivity to redox changes, (iii) increased sediment

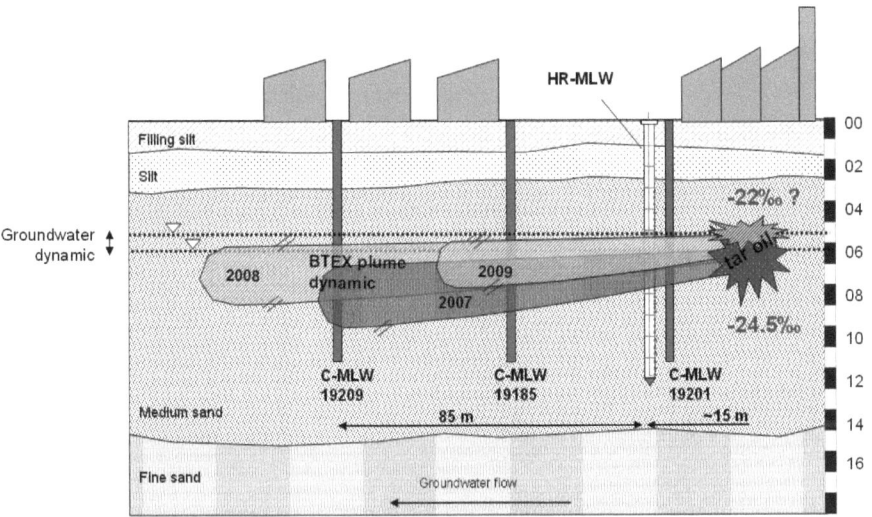

heterogeneity in the infrequently saturated vadose zone, (iv) temporal lateral influences from another source, as well as (v) the collapse of a bacterial population caused by bacterial competition, viral infection or protozoan grazing may be considered.

Figure 3.6: Schematic illustration of the field site highlighting the assumed plume dynamics during the years 2007 to 2009.

Without doubt, our data set demonstrates the spatiotemporal dynamic of a contaminant plume mainly composed of BTEX and naphthalene. Maximum contaminant concentrations varied by almost a factor of six within the observation period of four years at the spot of our

HR-MLW. The dynamics in contaminant concentrations along the vertical cross section at site of our HR-MLW, gave rise to a strong dynamic in plume lengths (Fig. 3.6). This assumption could be confirmed by an independent data set from repeated sampling of several adjacent conventional multi-level wells. Note, the much lower BTEX concentrations with these measurements are a result of the much lower vertical resolution of groundwater sampling, 50 - 100 cm instead of 3 - 10 cm at the HR-MLW (Fig. 3.5B). Sampling the C-MLW using a multi-channel suction pump delivered groundwater which integrated the chemical information of several cubic decimeters, leading to the underestimation of true contaminant concentration by a factor 5 and more (Fig. 3.5B; Anneser et al. 2008a, Anneser et al. 2008b).

The observed dynamics including the part time collapse of toluene degradation opens a new discussion on the reliability of NA processes, assessment of NA and enhanced NA as conservative long-term sanitation concepts. Enhancement of our knowledge will help to improve the predictability of plume development and to assess the reach of NA and further remediation strategies. These findings ask for a future collaboration for the further development of reactive transport models, now considering microbial growth, death and interactions in terms of functional redundancy and biodegradation.

3.4. References

Alfreider A, Krössbacher M, Psenner R (1997) Groundwater samples do not reflect bacterial densities and activity in surface systems. Water Research 31:882-840

Anneser B, Einsiedl F, Meckenstock RU, Richters L, Wisotzky F, Griebler C (2008a) High-resolution monitoring of biochemical gradients in a tar oil-contaminated aquifer. Applied Geochemistry 23:1715-1730

Anneser B, Pilloni G, Bayer A, Lueders T, Griebler C (2010) High Resolution Analysis of Contaminant Aquifer Sediments and Groundwater - What Can be learned in Terms of Natural Attenuation? Geomicrobiology Journal 27:130-142

Anneser B, Richters L, Griebler C (2008b) Application of high-resolution groundwater sampling in a tar oil-contaminated sandy aquifer. Studies on small-scale abiotic gradients. In: Candela L, Vadillo I, Elorza FJ (eds) Advances in Subsurface Pollution of Porous Media: Indicators, Processes and Modelling CRC press/Balkema, Taylor & Francis Group, London, UK, p 107-122

Bauer RD, Maloszewski P, Zhang Y, Meckenstock RU, Griebler C (2008) Mixing-controlled biodegradation in a toluene plume - Results from two-dimensional laboratory experiments. Journal of Contaminant Hydrology 96:150-168

Bauer RD, Rolle M, Bauer S, Eberhardt C, Grathwohl P, Kolditz O, Meckenstock RU, Griebler C (2009) Enhanced biodegradation by hydraulic heterogeneities in petroleum hydrocarbon plumes. Journal of Contaminant Hydrology 105:56-68

Berney M, Vital M, Hülshoff I, Weilenmann H-U, Egli T, Hammes F (2008) Rapid, cultivation-independent assessment of microbial viability in drinking water. Water Research 42:4010-4018

Brielmann H, Griebler C, Schmidt SI, Michel R, Lueders T (2009) Effects of thermal energy discharge on shallow groundwater ecosystems. FEMS Microbiology Ecology 68:273-286

Cozzarelli IM, Herman JS, Baedecker MJ, Fischer JM (1999) Geochemical heterogeneity of a gasoline-contaminated aquifer. Journal of Contaminant Hydrology 40:261-284

Elliot DR, Scholes JD, Thornton SF, Rizoulis A, Banwart SA, Rolfe SA (2010) Dynamic changes in microbial community structure and function in phenol-degrading mirocosms inoculated with cells from a contaminated aquifer. FEMS Microbiology Ecology 71:247-259

Griebler C (2001) Microbial ecology of subsurface ecosystems. In: Griebler C, Danielopol DL, Gibert J, Nachtnebel HP, Notenboom J (eds) Groundwater ecology - a tool for management of water resources. Office of Official Publications of the European Communities, Luxembourg, p 81-108

Griebler C, Lueders T (2008) Microbial biodiversity in groundwater ecosystems. Freshwater Biology 54:649-677

Hammes F, Berney M, Wang Y, Vital M, Köster O, Egli T (2008) Flow-cytometric total bacterial cell counts as a descriptive microbial parameter for drinking water treatment processes. Water Research 42:269-277

Hendrickx B, Dejonghe W, Boenne W, Brennerova M, Cernik M, Lederer T, Bucheli-Witschel M, Bastiaens L, Verstraete W, Top E, Diels L, Springael D (2005) Dynamics of an Oligotrophic Bacterial Aquifer Community during Contact with a Groundwater Plume Contaminated with Benzene, Toluene, Ethylbenzene and Xylenes: an In Situ Mesocosm Study. Applied and Environmental Microbiology 71:3815-3825

Kirchman DL (1993) Leucine incorporation as a measure of biomass production by heterotrophic bacteria. In: Kemp PF, Cole JJ, Sherr BF, Sherr EB (eds) Handbook of methods in aquatic microbial ecology. Lewis publisher, New York

Kirchman DL, Ducklow HW (1993) Estimating conversion factors for the thymidine and leucine methods for measuring bacterial production. In: Kemp PF, Cole JJ, Sherr BF, Sherr EB (eds) Handbook of methods in aquatic microbial ecology. Lewis publishers New York

Kisand V, Noges T (2004) Abiotic and biotic factors regulating dynamics of bacterioplankton in a large shallow lake. FEMS Microbiology Ecology 50:51-62

Kuhn TK, Hamonts K, Dijk JA, Kalka H, Stichler W, Springael D, Dejonghe W, Meckenstock RU (2009) Assessment of the Intrinsic Bioremediation Capacity of an Eutrophic River Sediment Polluted by Discharging Chlorinated Aliphatic Hydrocarbons: A Compound-Specific Isotope Approach. Environmental Science and Technology 43:5263-5269

Lewandowski G, Mortimer G (2004) Estimation of Anaerobic Biodegradation Rate Constants at MGP Sites. Groundwater 42:433-437

Mailloux BJ, Fuller ME (2003) Determination of In Situ Bacterial Growth Rates in Aquifers and Aquifer Sediments. Applied and Environmental Microbiology 69:3798-3808

Mayer KU, Benner SG, Frind EO, Thornton SF, Lerner DN (2001) Reactive transport modeling of processes controlling the distribution and natural attenuation of phenolic compounds in a deep sandstone aquifer. Journal of Contaminant Hydrology 53:341-368

McGuire JT, Long DT, Hyndman DW (2005) Analysis of recharge-induced geochemical change in a contaminated aquifer. Ground Water 43:518-530

McGuire JT, Long DT, Klug MJ, Haack SK, Hyndman DW (2002) Evaluating Behavior of Oxygen, Nitrate, and Sulfate during Recharge and Quantifying Reduction Rates in a Contaminated Aquifer. Environmental Science & Technology 36:2693-2700

McGuire JT, Smith EW, Long DT, Hyndman DW, Haack SK, Klug MJ, Velbel MA (2000) Temporal variations in parameters reflecting terminal-electron-accepting processes in an aquifer contaminated with waste fuel and chlorinated solvents. Chemical Geology 169:471-485

Meckenstock RU, Lueders T, Griebler C, Selesi D (2010) Microbial Hydrocarbon Degradation at Coal Gasification Plants. In: Timmis K (ed) Handbook of Hydrocarbon and Lipid Microbiology. Springer, Berlin-Heidelberg, p 2293-2312

Meckenstock RU, Morasch B, Griebler C, Richnow HH (2004) Stable isotope fractionation analysis as a tool to monitor biodegradation in contaminated aquifers. Journal of Contaminant Hydrology 75:215-255

Pilloni G (2011) Distribution and dynamics of contaminant degraders and microbial communitites in stationary and non-stationary contaminant plumes. Technische Universität, München

Pilloni G, von Netzer F, Engel M, Lueders T (2011) Electron acceptor-dependent identification of key anaerobic toluene degraders at a tar-oil-contaminated aquifer by Pyro-SIP. FEMS Microbiology Ecology:1-11

Prommer H, Barry DA, Davis GB (2002) Modelling of physical and reactive processes during biodegradation of a hydrocarbon plume under transient groundwater flow conditions. Journal of Contaminant Hydrology 59:113-131

Rees HC, Oswald SE, Banwart SA, Pickup R, Lerner DN (2007) Biodegradation Processes in a Laboratory-Scale Groundwater Contaminant Plume Assessed by Fluorescence Imaging and Microbial Analysis. Applied and Environmental Microbiology 73:3865-3876

Rein A, Bauer S, Dietrich P, Beyer C (2009) Influence of temporally variable groundwater flow conditions on point measurements and contaminant mass flux estimations. Journal of Contaminant Hydrology 108:118-133

Robinson C, Brovelli A, Barry DA, Li L (2009) Tidal influence on BTEX biodegradation in sandy coastal aquifers. Advances in Water Resources 32:16-28

Schäfer D, Schlenz B, Dahmke A (2004) Evaluation of exploration and monitoring methods for verification of natural attenuation using the virtual aquifer approach. Biodegradation 15:453-465

Sherwood Lollar B, Hirschhorn SK, Chartrand MMG, Lacrampe-Souloumc C (2007) An approach for assessing total instrumental uncertainity in compound-specific carbon isotope analysis: Implications for environmental remediation studies. Analytical Chemistry 79:3469-3475

Thullner M, Mauclaire L, Schroth MH, Kinzelbach W, Zeyer J (2002) Interaction between water flow and spatial distribution of microbial growth in a two-dimensional flow field in saturated porous media. Journal of Contaminant Hydrology 58:169-189

Tuxen N, Albrechtsen H-J, Bjerg PL (2006) Identification of a reactive degradation zone at a landfill leachate plume fringe using high resolution sampling and incubation techniques. Journal of Contaminant Hydrology 85:179-194

van Breukelen BM, Griffioen J, Röling WFM, van Verseveld HW (2004) Reactive transport modelling of biochemical processes and carbon isotope geochemistry inside a landfill leachate plume. Journal of Contaminant Hydrology 70:249-269

Van Breukelen BM, Griffion J (2004) Biogeochemical processes at the fringe of a landfill leachate pollution plume: potential for dissolved organic carbon, Fe(II), Mn(II), NH_4, and CH_4 oxidation. Journal of Contaminant Hydrology 73:181-205

Vencelides Z, Sracek O, Prommer H (2007) Modelling of iron cycling and its impact on the electron balance at a petroleum hydrocarbon contaminated site in Hnevice, Czech Republic. Journal of Contaminant Hydrology 89:270-294

Vroblesky DA, Chapelle FH (1994) Temporal and spatial changes of terminal electron-accepting processes in a petroleum hydrocarbon-contaminated aquifer and the significance for contaminant biodegradation. water resources research 30:1561-1570

Wilhartitz IC, Kirschner AKT, Stadler H, Herndl GJ, Dietzel M, Latal C, Mach R, Farnleitner AH (2009) Heterotrophic prokaryotic production in ultraoligotrophic alpine karst aquifers and ecological implications. FEMS Microbiology Ecology 68:287-299

Winderl C, Anneser B, Griebler C, Meckenstock RU, Lueders T (2008) Depth-resolved quantification of anaerobic toluene degraders and aquifer microbial community patterns in distinct redox zones of a tar oil contaminant plume. Applied and Environmental Microbiology 74:792-801

Yagi JM, Neuhauser EF, Ripp JA, Mauro DM, Madsen EL (2010) Subsurface ecosystem resilience: long-term attenuation of subsurface contaminants supports a dynamic microbial community. ISME Journal 4:131-143

4. Quantification and preservation of aquifer sediment bacteria – a multiple assay comparison

Quantification of the total bacterial number in aquatic sediment samples is challenging. We have developed and systematically evaluated a standard protocol for estimation of total bacterial abundance optimized for sandy sediments. Sample preparation including density gradient centrifugation and quantification via flow cytometry were combined. Moreover, the effects of sediment freezing (-20°C, liquid nitrogen, dry ice) on total cell numbers, the number of active cells derived from ATP concentration measurements, and 16S rDNA gene copies have been investigated. Initial freezing of sediments leads to a significant loss of cells counts in the range of 20 - 90%. Shock freezing using liquid nitrogen proved to be least destructive. Repeated freezing did not further decrease cell numbers significantly, pointing at a certain proportion of the bacterial community to be persistent to freezing and thawing stress. Apparently mostly the part of a community active at time of sampling is mainly lost, as indicated from the ATP measurements. Most interestingly, freezing unexpectedly lead to a strong decline in 16S rDNA gene copies. If it turns out true that the most active fraction of a bacterial community is lost after freezing, functionally and phylogenetically dominant species or groups within bacterial communities may be underrepresented after molecular analysis of sediment samples stored at frozen conditions. This point clearly needs further investigation.

4.1. Introduction

With groundwater filled aquifers, the terrestrial subsurface harbors the world's largest freshwater ecosystem (Danielopol et al. 2003). The biomass of prokaryotes in the subsurface domain accounts for about 6 - 40% of the earth's total microbial biomass, which is a minimum estimate, based on unconsolidated sediments representing only 20% of the terrestrial subsurface (Whitman et al. 1998). Most of these prokaryotes (Eubacteria and Archaea) are associated with sediment surfaces, forming micro colonies and biofilms, while only a small fraction (0.01 - 10%) are found suspended in the pore water (Alfreider et al. 1997, Griebler et al. 2002, Duhamel & Jacquet 2006, Chang et al. 2009). Consequently, the analysis of prokaryotes, in the following simply termed bacteria, requires reliable techniques for their preservation and quantification.

The first step in the quantification of sediment bacteria is the adequate preservation. Most commonly, water and sediment samples dedicated to direct counts are fixed with paraformaldehyde, formaldehyde, glutardialdehyde or a combination of metal ions and sodium azide (Günther et al. 2008). With other quantification approaches, such as DNA extraction and subsequent qPCR of 16S rDNA genes, sediments are routinely (shock) frozen and kept at ≤ -20°C till further processing (Madsen 2000). Especially freezing and thawing is known to cause destruction of a fraction of bacteria (Skogland et al. 1988, Ellison et al. 1991, Dodd et al. 1997, Dodd et al. 2007). To date, there is hardly any little information if this loss is significant and systematic and, if frozen sediment samples may serve as backup for total bacterial counts.

Quantification of bacteria is mainly based on direct counting, either via epifluorescence microscopy (Dayley & Hobbie 1975, Porter & Feig 1980, Schallenberg et al. 1989, Epstein & Rossel 1995, Kuwae & Hosokawa 1999) or flow cytometry (Vives-Rego et al. 2000, Ziglio et al. 2002, Gruden et al. 2004, Amalfitano & Fazi 2008). For statistical reasons a high number of cells per sample needs to be counted (Kirchman et al. 1982) which makes the microscopic approach very time-intensive and limits the number of samples processed in a study. In contrast, flow cytometry enables the researcher high throughput measurements. Both techniques are applied routinely with liquid samples, but it is still challenging to quantify bacteria attached to sediments.

With respect to sediment bacteria, additional difficulties occur. In detail, the successful detachment of cells from particles minimizing cell destruction and getting rid of organic and inorganic particles others than bacteria. Removal of cells is mainly accomplished by homogenization and sonication, while removal of unspecific particles is mainly achieved via sample dilution (Gough & Stahl 2003) and density gradient centrifugation (DGC) (Bakken & Lindahl 1995, Mayr et al. 1999, Courtois et al. 2001, Amalfitano & Fazi 2008). It has been shown that the success of a detachment procedure strongly depends on specific sediment properties and thus the optimal protocol has to be determined individually for each sediment sample type (Buesing & Gessner 2002). Based on total DNA extraction, and thus independent from a careful and quantitative detachment of cells from sediment grains, is qPCR (Witzingerode et al. 1997, Bach et al. 2002) and ATP analysis (Hammes et al. 2008). These approaches are accompanied to other individual potential uncertainties, including the incomplete lysis of cells and inhibitory effects of humics as well as other compounds during PCR and ATP analysis (Witzingerode et al. 1997). Last but not least, indirect methods often suffer from inadequate standardization and conversion factors. Moreover, cell numbers gained from ATP measurements represent the active microbial biomass which may considerably deviate from the total bacterial biomass.

Here we report on an optimized protocol for a fast and direct quantification of prokaryotic cells from sandy aquifer sediments in a flow cytometer and named it in the following "standard protocol". In comparison with alternative indirect approaches for microbial quantification, *i.e.* qPCR and ATP, the quantitative effects of sample storage, taking into account repeated freezing and thawing cycles, are examined.

4.2. Materials and methods

4.2.1. Sediment samples

Most experiments, during development of a standard protocol for the quantification of aquifer sediment bacteria, were conducted with quaternary sandy sediment (grain diameter 0.063 - 4 mm) which was percolated with groundwater from a quaternary aquifer in a barrel for several months. At some occasions also sterile quartz sand (grain diameter: 200 - 300 µm) was used as model sediment (see below). Later in the study, fresh sediment samples from a sandy porous aquifer at a former gas works site in Düsseldorf-Flingern (Germany) were collected by means of drilling. Along a vertical cross section through a contaminant plume of aromatic hydrocarbons, sediment subsamples originated from 6.0 to 10.4 m below land surface (bls) were analyzed. A detailed description of the sampling site is given elsewhere (Anneser et al. 2008, Anneser et al. 2010).

4.2.2. Culture bacteria

For the purpose of standardization and evaluation of recovery, cultures of *Pseudomonas putida F1* and *Aromatoleum aromaticum* EbN1 were grown in an oxic or anoxic carbonate-buffered freshwater mineral medium, respectively. Cells were harvested at the end of their exponential growth phase by centrifugation. In case of *P. putida*, the pellet was resolved in PBS buffer (Dulbecco, 1x, Ca^+, Mg^{2+}, Biochrom AG), fixed in 2 mL 2.5% glutardialdehyde and stored at +4°C for further use. Use of *A. aromaticum* as standard in qPCR is described in more detail below.

4.2.3. Densitiy gradient centrifugation (DGC)

Since high background of unspecifically stained cells constituted the most serious problem with the quantification of sediment bacteria, the reduction of these small inorganic bacterialike particles via DGC was the first step evaluated. 0.5 mL of fresh sediment was placed into a 2 mL Eppendorf tube and fixed with 0.5 mL formaldehyde (37%) for 10 min (Griebler et al. 2001). Thereafter, 1.5 mL of particle free PBS buffer was added, subsequently the sample was briefly mixed by hand and spun down at 15,000 g for 10 min in an Eppendorf centrifuge. Again 1.5 mL of PBS buffer was added and the tube was placed into a swing mill (12 Hz, 3 min; H. Eisenmann pers. commun.) to dislodge the attached prokaryotic cells. The supernatant (1.5 mL) having room temperature was placed onto 5 mL of a cold Nycodenz solution (Progen Biotechnik GmbH) at a final concentration of 1.3 g mL^{-1} at pH 8 (Lindahl & Bakken 1995) in a 10-mL-centrifuge tube. DGC was done in a Centricon ultra centrifuge (T-2190, TST 41.14 swing rotor) and particles were spun down at 15,500 g for 1 h at 4°C. After centrifugation subsamples were collected in 1 mL portions by pipetting starting at the top (tubes containing 6.5 mL in total).

To qualitatively and quantitatively verify the centrifugation step, total bacterial numbers of triplicate samples were determined microscopically (see below) in all sample fractions. As a first test, cells of a *Pseudomonas* culture were diluted in 1.5 mL PBS aliquots to a final concentration of ca. 10^7 cells mL^{-1}. The aliquots were applied on 5 mL of cold Nycodenz solution and centrifuged as described above. The total recovery from the different sample fractions after DGC was determined in comparison to the total microscopic counts of non-centrifuged control samples. A similar test was run with quartz sand samples spiked with *Pseudomonas* cells. To estimate the amount of bacterial cells spun down together with particles, the pellet in the centrifugation tubes were exemplarily resolved in 1.5 mL PBS buffer, treated in the swing mill and applied once again on cold Nycodenz for repeated centrifugation. Finally, real sediment samples (quaternary sandy aquifer sediment) were treated in the swing mill and the supernatant was placed onto the Nycodenz solution.

4.2.4. Flow cytometry

Sample aliquots (mostly 100 µL) were stained with SYBR Green I (5 µL mL^{-1} working solution, see below) and incubated for 10 min at 4°C in the dark. As an internal standard, Trucount beads (Becton Dickinson) dissolved in fresh 0.2 µm-filtrated PBS buffer were used. For practical reasons and economic consideration, one Trucount tube was amended with 4 mL of PBS buffer and split into four aliquots. Sample analysis was performed with a LSR II flow cytometer (488 nm and 633 nm laser, Becton Dickinson). The instrument settings were as follows: forward scatter (FSC) 350 mV, side scatter (SSC) 300 - 370 mV, bandpass filter 530 nm (B530) 500 - 580 mV, and bandpass filter 610 nm (B610) 650 mV depending on optimal separation of *Pseudomonas* cells and standard beads. All parameters were collected as logarithmic signals. The bacterial counting gate was initially set and repeatedly checked with a *P. putida* culture comparing green (B530) and red (B610) fluorescence and further adjusted to separate the bacterial populations from background particles. To minimize background noise, the signal threshold was adjusted to 200 mV. All analyses were performed at a minimum flow rate (about 10 µL min^{-1}). Samples were always measured in triplicates and the BD FACSDiva software package (Becton Dickinson) was used for data analysis. In order to compare the results obtained by flow cytometry with results from microscopy, aliquots of the same samples were counted in parallel.

Calculation of the total number of bacteria per mL sediment (N_{bac}), was derived through the experiments, according to:

$$N_{bac} = \frac{bac_{counted} \times beads_{total} \times V_{Fractions}}{beads_{counted} \times V_{sample} \times V_{sediment}} \times 1.43 \times 1.28 \times 1.1 \quad (3)$$

Where $bac_{counted}$ = number of events in the gate for the bacterial cells, $beads_{counted}$ = number of events in the gate for the beads, $beads_{total}$ = the total number of beads in the Trucount tube, $V_{Fractions}$ = volume of the measured fractions [mL], V_{sample} = volume of the sample taken for the flow cytometer [mL], $V_{sediment}$ = volume of used sediment [mL], **1.43** = factor for the release efficiency (see results), **1.25** = factor for the loss of cells during the DGC not found in the selected fractions 2 & 3 (see results), and **1.1** = corrects for the sediment pore water not transferred after dislodgment.

4.2.5. Microscopic counts

Subsamples after DGC (0.5 to 2 mL) were diluted in particle free PBS buffer to a final volume of 5 mL, stained with 25 µL working solution of SYBR Green I and incubated for 10 min at 4°C in the dark. The samples were briefly mixed and filtered onto 0.2 µm black polycarbonate filters (Millipore). Subsequently filters were washed once with 5 mL PBS buffer, removed from the filtration tunnel and embedded in fluorescent oil on a microscope slide. Cells were counted with a Zeiss Axiolab microscope at 1000x magnification (filter set: Zeiss, Ex 450 - 490 nm, FT 505 nm, LP 520 nm) using a counting grid for quantifications. For statistical reasons, a minimum of 200 - 400 cells and not less then 10 observation fields were examined (Kirchman et al. 1982). The total number of bacterial cells (N_{bac}) was calculated as follows:

$$N_{bac} = \frac{S \times 10^6 \times n}{s \times V} \quad (4)$$

Where **S** = filter area [mm^2], **n** = mean of bacterial numbers per working field, **s** = area of working field [µm^2], **V** = volume of water filtered [mL].

4.2.6. Sediment preservation tests

Different fixation protocols were tested for sediment samples in relation to the quantity and quality of staining applying different fluorescent dyes and ways of preservation. Aliquots of 0.5 mL of fresh sediment were fixed in (1) 0.5 mL of 37% formaldehyde, (2) 1.5 mL of 4% formaldehyde, or (3) 1.5 mL of 2.5% glutardialdehyde, and stored at 4°C in the dark (Griebler et al. 1997, Bast 2001, Böckelmann et al. 2003). Additionally, different sediment aliquots were fixed in 1.5 mL of 4% paraformaldehyde, incubated for three days at 4°C in the dark, then washed twice with 1 mL of PBS and spun down at 4000 *g* for 30 min. Subsequently these sediment samples were resolved and stored in a mixture of 0.75 mL PBS buffer and 0.75 mL ethanol absolute (Merck) at 4°C in the dark (Bachoon et al. 2001).

Preservation was conducted by placing fresh sediment aliquots at -20°C, on dry-ice, or into liquid nitrogen. All frozen sediments were stored at -20°C until further processing. Thawing of frozen samples generally took place at room temperature, occasionally putting the frozen sediment directly in PBS buffer. For reasons of comparison, numerous aliquots underwent repeated (two to three times) freezing and thawing cycles. All sediment aliquots which have been frozen prior to fixation were treated with 1.5 mL 2.5% glutardialdehyde after thawing.

4.2.7. Detachment, loss and damage of cells

Dislodgement of bacterial cells from sediment particles was tested comparatively (1) in an ultrasonic bath (Branson digital sonifier 450) at 20% amplitude from 20 to 360 s with pulsed and continuous sonication, (2) in a swing mill (MM 200, Retsch) at 10, 20 and

30 Hz and (3) using a test tube shaker (Thermomixer comfort, Eppendorf) at 1400 rpm. These treatments were tested with culture bacteria amended to sterile aquifer sediments or quartz sand to evaluate cell damage and recovery. Some sample aliquots received the addition of the surface active detergent sodium pyrophosphate (PPI, 100 mM), prior to the mechanical treatment (Griebler et al. 2001). Different settings of ultrasonication were further tested to properly disaggregate cell pellets of culture bacteria after centrifugation steps without pronounced cell damage or loss. Cell aggregation and possible damage was checked microscopically. To test the efficiency of cell release from sediment samples treated in the swing mill, the supernatant was collected and replaced by new PBS buffer. The sediment sample was homogenized again. This procedure was repeated three times and the subsequently collected supernatants were further processed for DGC as described above.

4.2.8. Staining of prokaryotes

For staining, a number of fluorescent dyes were tested with epifluorescence microscopy at various concentrations; 4',6-diaminido-2-phenylindole (DAPI, Diagnostica Merck) at 0.01 to 0.5 mg mL^{-1}, CYTO 59 (5 mM, Molecular Probes) at 0.1 µL mL^{-1} and 0.2 µL mL^{-1}, acridine orange (Sigma) at 4 µg mL^{-1}, and SYBR Green I (stock 10 000x concentrated, Invitrogen; working solution 1:10 with PBS buffer) from 0.5 to 20 µL mL^{-1} final concentration. Based on the results from microscopic test, only SYBR Green I (3 µL mL^{-1} and 6 µL mL^{-1}) and CYTO 59 (0.6 µL mL^{-1}) were further evaluated via flow cytometry.

In addition, 0.5 mL aliquots of sterile sediment diluted in 1.5 mL PBS buffer were stained with SYBR Green I (3 µL mL^{-1}) after treatment in the swing mill (20 Hz, 3 min). Some of these sample aliquots were centrifuged at 10,000 g for 10 min, then washed with PBS after staining, and compared to sterile sediment aliquots treated the same way but without staining. These samples were examined by flow cytometry to check for unspecific staining and de novo generation of small particles due to the detachment procedure.

4.2.9. Adenosine-triphosphate (ATP)

Total ATP was determined as described in Velten et al. (2007) using BacTiter-Glo^{TM} reagent (Promega) and a CellScan luminometer (Taurus Instruments). In detail, 0.2 mL sediment with 60 µL ATP-free Milli-Q water (MQ) was heated to 37°C in a Thermomixer (Thermomixer comfort, Eppendorf). 100 µL of the 37°C-warm BacTiter-Glo^{TM} reagent was added and mixed briefly in the Thermomixer at 1400 rpm for 2.5 min. After addition of 900 µL preheated ATP-free MQ with subsequent short centrifugation of 20 s by maximal speed to spin down the particles, the supernatant was transferred into a cuvette. Luminescence was measured integrating over 10 s. Values were then converted to ATP concentration using a calibration curve by using known ATP concentrations (Roche) diluted in ATP-free MQ and performed at 37°C. Assuming a mean concentration of 1.75×10^{-10} nmol ATP per cell (Hammes et al. 2010) the number of active cells was calculated.

4.2.10. Nucleic acid extraction and quantitative PCR

DNA from sediment samples were extracted in triplicates as described in Lueders et al. (2004). The investigated sediment samples were ~0.7 g fresh, frozen in liquid nitrogen, frozen at -20°C, refrozen at -20°C after thawing or thawed in 1x PBS buffer, always without any fixative. The DNA was stored at -20°C and the amount of extracted DNA was measured with PICO Green (Invitrogen) followed by quantitative PCR (qPCR). With some modifications qPCR was performed as previously described in Winderl et al. (2008). Only 40 cycles were needed with denaturation and annealing time of 30 s each. The primer set 519f/907r (Muyzer et al. 1995, Biddle et al. 2008) was used to quantify the extracted amount of bacterial 16S rDNA.

To correct for potentially distinct amplification/detection efficiencies of the qPCR assay and also for putative extraction/detection efficiencies of our general workflow, defined biomass amendments were evaluated for the Düsseldorf-Flingern aquifer sediments. For this, sediments were sterilized over night at 180°C to eliminate intrinsic nucleic acids. Then, sediments were re-wetted with sterile water and amended with defined cell numbers of a freshly grown, SYBR Green-counted liquid culture of "*A. aromaticum*" EbN1 between 8×10^5 and 8×10^6 cells g^{-1} wwt of sediment. Care was taken to adjust the sediments to original water content. Strain EbN1 carries four *rrn*-operons per genome (Rabus & Widdel 1995). Nucleic acids were re-extracted and quantified as described above. From the detected vs. expected gene quantities, a correction factor for 16S gene counts directly obtained from Flingern sediments was inferred. Negative controls showed that heat sterilization destroyed over 99.99% of initial 16S rDNA gene counts.

4.2.11. Statistics

For statistical evaluation of the data, the statistic package in SIGMA Plot 11 and the software R (version 2.9.1) were used. In SIGMA Plot, the student's t-test or the Mann-Whitney Rank test was applied for comparison of two groups of values. To compare multiple groups, the one-way ANOVA was used. In R, the two-sample t-test with unequal variances to test statistical significance was applied. Differences were considered to be significant at a confidence level of $p < 0.05$ in all tests.

4.3. Results and discussion

Total number of bacteria and prokaryotic biomass in sediment is a crucial parameter to understand an ecosystem. A close link has been shown between total number of bacteria and sediment grain size, oxygen concentration, organic matter content, perturbation and grazing (Jones 1974, Meyer-Reil 1993, Lucchesi & Santangelo 1997, Griebler et al. 2002). Similar relationships may be assumed for groundwater and aquifer sediments (Griebler & Lueders 2009). To date, only a minute fraction of the subsurface has been sampled and investigated and data on the prokaryotic abundance and biomass is extremely scarce, especially when compared to data available from surface ecosystems. Without doubt, this is partly caused by the inaccessibility of the subsurface and the enormous costs associated with drilling activities. However, generation of data is also hampered by time-consuming methods applied for the assessment of total bacterial counts in sediment. Fast and reliable methods are thus urgently needed for the assessment of total bacterial numbers in aquifer sediments. Moreover, the effects of sample preservation by fixatives or freezing with respect to loss of cells as well as rDNA copies need careful consideration.

4.3.1. Density gradient separation of bacteria and inorganic particles

DGC of cells recovered from quartz sand spiked with *P. putida* or detached from river bed sediment and sediment from Düsseldorf-Flingern freshly fixed with glutardialdehyde resulted in the bulk of bacterial cells in fraction 2 and 3 (Fig. 4.1), with the major portion of cells contained in fraction 2. A more detailed picture on the distribution of cells along the Nycodenz density gradient is given in table 4.1.

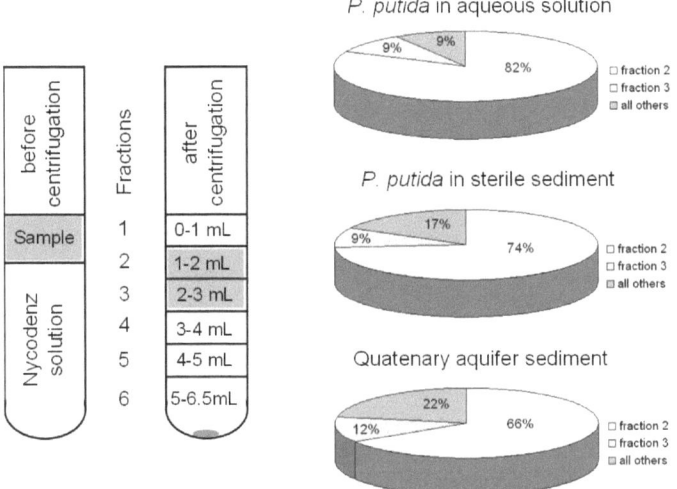

Figure 4.1: Distribution of bacterial cells in different fractions after density gradient centrifugation exemplarily shown for different types of sediment samples.

Because a series of natural samples, including spiked quartz sand, aquifer and river bed sediments, examined revealed 75% to 85% of all cells distributed in the fractions 2 and 3 (data not shown), we set a correction factor of 1.25 for bacteria in natural sediment samples to correct the loss of cells in the not used fractions. Similarly, Lindahl and Bakken (1995) found more than 95% of a pure culture of bacteria and about 89% of soil bacteria in the cushion (comparable to fraction 2 in our approach) of the Nycodenz after DGC and 87% recovery of cells are reported by Priemé et al. (1996). Up to 30% of the cells may be found in the pellet (Kallmeyer et al. 2008), which results from cells remaining sorbed to sediment particles or due to co-sedimentation. In our experiments, only 4 - 9.5% of the cells were found in the final fraction which also contained the pellet. Additional to counting the cells via flow cytometry after our standard protocol procedure, the pellets have been checked microscopically (data not shown) revealing similar results.

Table 4.1: Distribution of the cells in the Nycodenz-gradient determined by flow cytometry.

Fraction	mL	P. putida + sterile sediment	P. putida	P. putida + Quartz sand	Groundwater sediment
1	0 – 1	0.5 ± 1%	1 ± 1%	3.5 ± 6.5%	1 ± 0.5%
2	1 – 2	74 ± 4%	81.5 ± 9.5%	68 ± 15.5%	66 ± 10%
3	2 – 3	9 ± 5%	9 ± 7.5%	12 ± 7.2%	12 ± 6.5%
4	3 – 4	3.5 ± 2%	1.5 ± 1%	8.5 ± 4%	5.5 ± 4%
5	4 – 5	6 ± 2%	1.5 ± 1%	3.5 ± 0.2%	6 ± 2%
6	5 – 6.5	7 ± 0.5%	5.5 ± 8%	4 ± 2.2%	9.5 ± 8%

The qualitative improvement of sample preparation, *i.e.* the reliable separation of non-bacterial particles from cells, is highlighted in Fig. 4.2. In charts A and B the tremendous reduction of the huge particle load is obvious. Moreover, spiking of sterile sediment with *P. putida* cells confirmed the reliable separation of bacterial cells from other particles (Fig. 4.2C). Fig. 4.3D underlines the unspecific staining effect, when adding the fluorescence dye to sterile, organic carbon free (muffled) sediment (Fig. 4.2D and 4.2E).

Unspecific staining is an effect which makes it very difficult to separate bacterial cells from unspecifically stained particles in the microscope. The addition of PPI significantly increased the degree of non-specific background staining. Non-specific staining of sediment is well-known from many studies for nearly all kinds of DNA-specific dyes (Epstein & Rossel 1995, Nunan et al. 2003). In a more recent study by Klauth et al. (2004) different green fluorescent nucleic acid dyes have been compared focusing on non-specific staining of sediments. SYTOX Green exhibited a significantly lower background staining effect to sediments, but the results varied broadly, depending on the different types of soils (Klauth et al. 2004).

Plotting the green (B530) versus the red fluorescence (B610) allowed a reliable separation of bacterial cells and particles in sediment (Fig. 4.2). This was also shown for water samples (Hammes & Egli 2005).

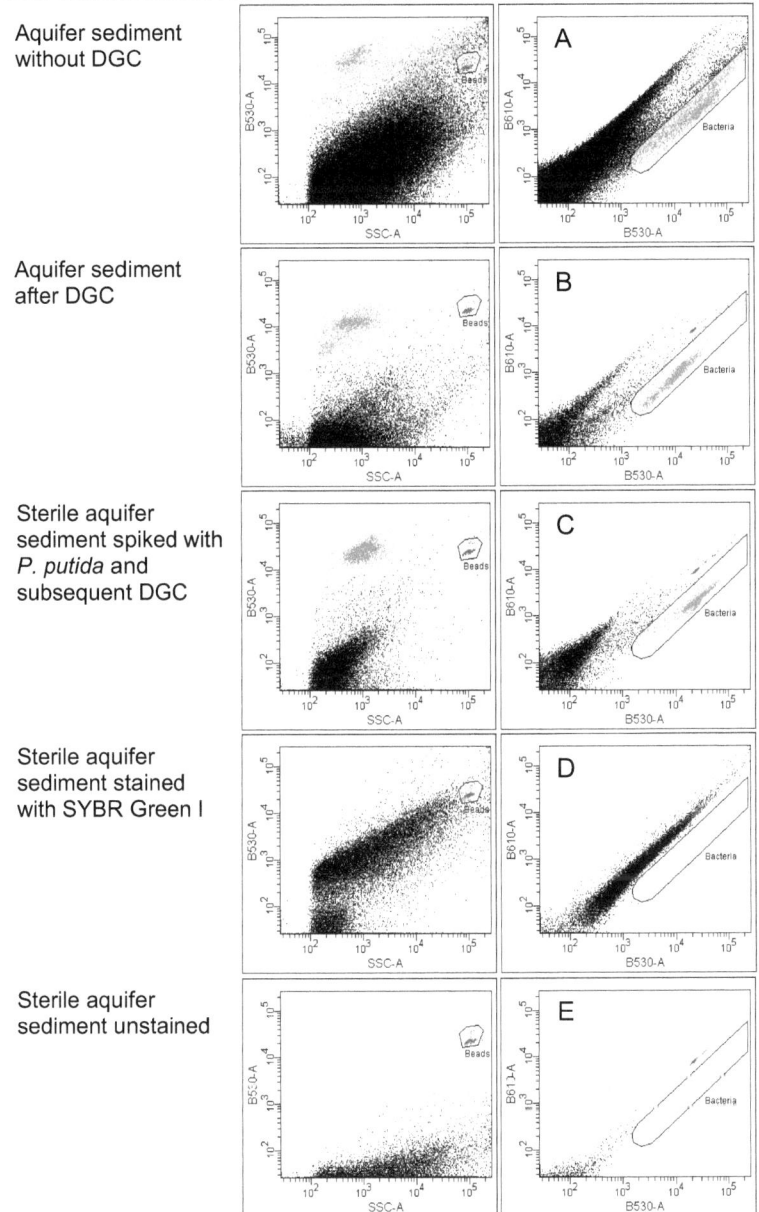

Figure 4.2: Flow cytometry data for (A) aquifer sediment processed without DGC, (B) treated following our evaluated standard protocol, (C) sterile sediment spiked with *P. putida* and subsequent DGC, (D) sterile sediment stained with SYBR Green I, and (E) unstained sterile sediment. All samples received the same amount of beads as internal standard.

4.3.2. Evaluation of different fixatives

The comparison of different fixatives and dyes at varying concentrations showed highest cell numbers, the best fluorescence yield and minimum background staining of particles when fixed with 2.5% glutardialdehyde and stained with SYBR Green I. This was evident for water and sediment samples. Fig. 4.3 shows the difference in total counts of SYBR Green I stained cells subsequent to various forms of fixation. Glutardialdehyde fixed samples revealed significantly higher cell numbers for microscopic (One Way ANOVA, P = <0.001 to 0.002) and flow cytometric counts (One Way ANOVA, P = 0.001 to 0.005).

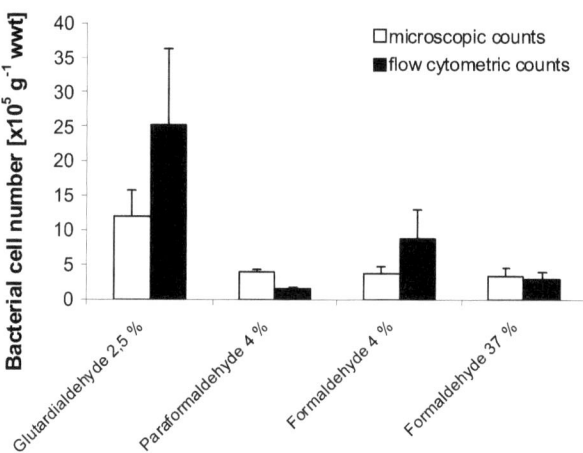

Figure 4.3: Total number of bacteria determined by flow cytometry and epifluorescence microscopy following different fixation treatments. Values are triplicates + SD.

It was shown, e.g. for marine water samples, that fluorescence was higher with unfixed living cells when compared to fixation with formaldehyde (Lebaron et al. 1998). This approach, however, has some limitations. First, samples without any fixation need to be cooled and processes within a few hours of collection (Ziglio et al. 2002). Fixation allows storage of samples for several weeks to months (Davey & Kell 1996). Even storage at 4°C may lead to significant cell loss, as it was shown for sediment samples (Duhamel & Jacquet 2006). Furthermore, fixation with aldehydes leads to denaturation of proteins and gives the cells robustness which is of great advantage during the process of dislodgement from sediment particles (Griebler et al. 2001).

When analyzing water, sample preservation with formaldehyde or paraformaldehyde is the most common method (Lebaron et al. 1998, Joachimsthal et al. 2003). However, a better performance of glutardialdehyde fixation when compared with formaldehyde and paraformaldehyde has been reported also by Lunau et al. (2005) and Martens-Habbena & Sass (2006), respectively. These and our results are in contradiction to the findings of Duhamel and Jacquet who obtained higher bacterial counts with sediment samples when fixed with formaldehyde than with glutardialdehyde (Duhamel & Jacquet 2006).

4.3.3. Efficiency of cell dislodgement and recovery

Homogenization (Dale 1974, Meyer-Reil et al. 1978, Dye 1983), ultrasonication (DeFlaun & Mayer 1983, Ellery & Schleyer 1984, Schallenberg et al. 1989, Epstein & Rossel 1995, Ward & Johnson 1996, Fischer et al. 1996), and the combination of both (Velji & Albright 1993, Weinbauer et al. 1998) have been found most efficient in the past. Homogenization separates bacteria from sediment particles by mechanical forces. Ultrasonication causes separation of bacteria and sediment particles as well as aggregation of bacteria by vibration of individual particles. Both techniques may also be applied to destroy or digest cells (Ellery & Schleyer 1984, Lindahl 1996). Consequently, the intensity and duration of these treatments are crucial. Here, homogenization was used to dislodge cells from sediment particles and brief ultrasonication was applied to disaggregate cells after DGC (see below). Treatment of the sediment samples in an ultrasonic bath was not further considered for cell detachment as it was found to produce a huge amount of small inorganic particles in preliminary experiments (data not shown).

To test the destructive force and the cell recovery efficiency related to homogenization by means of a mini shaker and a swing mill at different settings, sterile quartz sediment was spiked with a known number of cells from a *P. putida* culture. Processing sediment aliquots in the swing mill at 20 Hz for 3 min resulted in best recovery, while mixing on a shaker at 1400 Hz and 10 Hz for 3 min each resulted in significantly lower cell numbers (t-test, $p < 0.01$). Similarly, the swing mill treatment at higher intensity (30 Hz) exhibited significantly different, lower cell counts (t-test, $p < 0.05$), indicating cell disruption (Fig. 4.4).

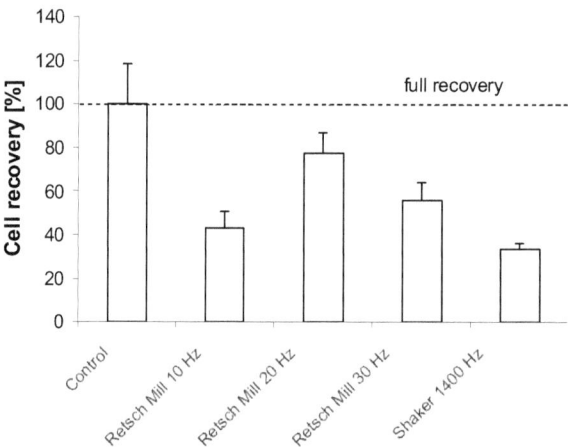

Figure 4.4: Recovery of culture bacteria (*P. putida*) spiked to sterile quartz sand subsequent to different homogenisation treatments, *i.e.* in a swing mill at different intensities or in a mini shaker. The number of cells spiked served as the control. Values are triplicates + SD.

Much research has compared procedures for detachment of bacterial cells from sediment particles. Buesing and Gessner (2002), for example, systematically compared ultrasonic bath, ultrasonic probe, homogenizer and blender for varying times of treatment of sediment and found the mechanical removal to be more efficient than ultrasonication, supporting our findings. However, a few other studies experienced the opposite (Ellery & Schleyer 1984). Mermillod-Blondin and coworkers (2001) compared ultrasonic bath against ultrasonic narrow tip generator and the latter was found to be more effective and less damaging to cells, which has been confirmed by Ellery and Schleyer (1984) and Epstein & Rossel (1995). Even apparently minor differences (e.g. instrument settings, power) in detachment procedures with the same basic approach and equipment can lead to significantly different results (Epstein & Rossel 1995, Buesing & Gessner 2002). Additionally to the time of sonification, size of the sample can also play an important role (Epstein & Rossel 1995). The addition of surface active detergents like PPI, Tween80 or methanol was shown to potentially increase the removal efficiency (Bakken & Lindahl 1995, Epstein & Rossel 1995), a treatment which was not tested during this study. Surface active reagents may interfere with cell counting (Velji & Albright 1993, Griebler 2001). Supplementary application of enzymatic treatment for the degradation of extracellular polymeric substances was used by Böckelmann and coworkers (2003) and Kallmeyer and coworkers (2008).

Testing the efficiency of cell removal applying the evaluated standard protocol repeated times to aquifer sediment samples, showed that only 70 ± 5% of the bacteria were released in the first treatment, while the following repetitions disclosed significantly lower numbers of bacteria; *i.e.* 17 ± 2% in the second, 8 ± 1% in the third and 5 ± 2% in the fourth run to calculate the total bacterial number. A higher release efficiency of 90% have been reported for sandy sediments applying a comparable detachment protocol (Griebler et al. 2001). However, much lower cell recovery of 25% and less have been reported (Lindahl 1996, Mayr et al. 1999). The success in detachment may vary with species and strains (Lindahl & Bakken 1995). One explanation can be the age of the bacterial cells, the "older" bacteria are the more adhesion to the particles they can have (Priemé et al. 1996). Without doubt, to obtain an optimized detachment protocol a compromise has to be found between detachment efficiency and cell damage. Based on the different removal efficiencies, various correction factors have been established (Ellery & Schleyer 1984, Griebler 2001). We set a correction factor of 1.43 in accordance to the 70% removal efficiency during a one step treatment with aquifer sediments. Approaches alternative to cell dislodgment is to simply correct for potential masking effects (Schallenberg et al. 1989) or to work with highly diluted samples to reduce the load of inorganic particles (Gough & Stahl 2003). Boenigk (2004) tried to reduce the number of interfering particles in sediments containing clay and fine silt by partial dissolution with hydrofluoric acid, but increased background fluorescence.

4.3.4. Comparison of different staining solutions

With microscopic counts, the best results for SYBR Green I stained samples were obtained at a final dye concentration of 0.8 - 1 µL mL^{-1}. At lower concentrations, part of the bacteria was not stained sufficiently, while higher concentrations resulted in an increase of background fluorescence. For flow cytometric analyses, there was no significant difference between the various SYBR Green I concentrations tested; consequently a final concentration of 0.3 µl mL^{-1} was applied. Comparing a number of fluorescence dyes, such as DAPI, CYTO 59 and Acridine Orange, SYBR Green I stained samples revealed the best results with epifluorescence microscopy. For flow cytometric analysis, only SYBR Green I and CYTO 59 have been tested, with SYBR Green I being most appropriate. Several independent studies support the finding that cells stained by green fluorescent dyes are easier to distinguish from non-specifically stained particles compared to other dyes such as DAPI (Lebaron et al. 1998, Weinbauer et al. 1998, Griebler et al. 2001, Gruden et al. 2003).

4.3.5. Microscopy vs. flow cytometry

When staining samples subsequent to DGC with SYBR Green I, microscopic observations showed bright greenish cells besides many of greenish and yellowish particles with just a moderate fluorescence. Only the green and bright fluorescing objects with the shape and size of bacteria were considered bacterial cells, leading to cell numbers slightly lower, but not statistically different as obtained by flow cytometry. A similar result was obtained when comparing flow cytometric and microscopic counts of samples from a bacterial batch culture of *P. putida* (Tab. 4.2). This correspondences with an earlier study comparing also both methods (Gruden et al. 2003).

Table 4.2: Total numbers of culture bacteria (*P. putida*) and sediment bacteria, as determined by flow cytometry and direct microscopic counting.

Counting method	*P. putida* [mL^{-1}]	Sediment bacteria [mL^{-1}]
Microscopy	$1.26 \times 10^8 \pm 2.25 \times 10^7$	$1.20 \times 10^6 \pm 2.25 \times 10^5$
Flow cytometry	$1.92 \times 10^8 \pm 3.45 \times 10^7$	$1.98 \times 10^6 \pm 2.74 \times 10^6$

An advantage of microscopic counts is that the samples can be observed directly, so particles can be assessed by taking into account their size and shape. However, the assessment of fluorescing objects based on their fluorescence intensity relies on the counter's subjective perception that has to decide to which extent the fluorescing particles are assessed as bacterial cells. Sediment samples are generally characterized by a high content of small inorganic particles, which maybe confused for bacterial cells, giving a range of error in cell count. Furthermore, human perception varies considerably among different individuals (Joachimsthal et al. 2003).

One advantage of flow cytometry is that fluorescing particles are only one time lighted when they pass the laser beam so fading is not a problem. The flow cytometer is able to exactly measure fluorescence intensities. Therefore, it can be clearly defined which particles are considered as bacteria based on their fluorescence and/or scattering properties. On the other hand, there is a certain range of error in data analysis. Setting the frames in the plots (see Fig. 4.2) contains uncertainties. The decision what to include in the frame relies at least partly on human judgment. Flow cytometry allows a far more rapid analysis of the samples, reduces the time needed for sample preparation as well as it constitutes an efficient tool for quantification of microorganisms in sediment samples.

For water flow cytometry is a routine methods to analyze cell counts as well as further parameters like cell size, cell structure, nucleic acid content and viability (Berney et al. 2008, Hammes et al. 2008, Hammes et al. 2010). Especially the separation of high and low nucleic acid bacteria is of ecological relevance (Gasol et al. 1999, Wang et al. 2009). This step still needs to be done for sediment samples in the future.

4.3.6. Evaluated „standard protocol"

Based on the improvements of the various steps in the quantification of sediment bacteria the following standard protocol was deduced: 0.5 cm^3 of fresh sediment is fixed with 1 mL of 2.5% glutardialdehyde and stored at 4°C in the dark until further preparations. After removal of the glutardialdehyde by centrifugation and discarding the supernatant, the sample is amended with 1.5 mL PBS and treated in a swing mill for 3 min at 20 Hz. The supernatant is collected and layered on 5 mL cold Nycodenz solution in a 10-mL-centrifuge tube and centrifuged as described earlier. Millilitre 2 and 3, the fraction with the main part of bacteria, are collected, of which an aliquot is subsampled and diluted in 0.2 µm-filtrated PBS (1:10 - 1:100) depending on the expected cell number of the sample. The samples are prepared for flow cytometric counts and counted as described above. This standard protocol was applied for investigating the influence of storage by freezing.

4.3.7. Influence of sample storage

The handling of fresh sediment samples directly after sampling has a pronounced influence on later quantification of bacteria. There is, for example, a significant difference between freshly processed samples and samples processed after freezing. A systematic comparison was conducted to qualitatively and quantitatively evaluate loss of bacterial cells, change of bacterial 16S rDNA copies, and ATP concentration subsequent to sample preservation by different ways of freezing. In all cases a significant reduction in cell counts have been observed when comparing to freshly processed aquifer sediment (Fig. 4.5A). Similarly, freezing led to a pronounced loss of 16S rDNA gene copies. This result is of high relevance, as freezing is the standard treatment for conserving environmental samples for later DNA extraction and molecular analysis. However, due to the high standard deviation of the replicates the differences were not significant (Fig. 4.5A). Comparing the loss of total cells, active cells and gene copies after freezing at -20°C and thawing at room temperature, with and without PBS buffer, the most pronounced differences to the freshly analysed samples was shown for ATP concentrations (active cells), followed by the gene copies and the number of total cells (Fig. 4.5B).

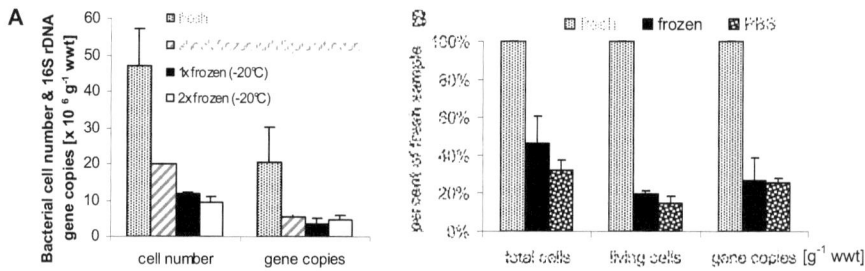

Figure 4.5: (A) Effect of different ways of freezing on the total number of bacterial cells and 16S rDNA gene copies. (B) Effect of freezing and thawing on the total number of bacterial cells, the number of living cells (derived from ATP measurements) and 16S rDNA gene copies. Freshly processed samples served as controls. Values are triplicates + SD.

There are studies reporting on sediment fixation after freezing at -20°C (Martens-Habbena & Sass 2006, Kallmeyer et al. 2008), although a loss in cell numbers due to freezing have been repeatedly observed (Helton et al. 2006). The increased effect of freezing to copy numbers of the 16S rDNA genes is most probably related to the loss of DNA during extraction, even when considering bacterial cells can contain more than one copy (Fogel et al. 1999, Klappenbach et al. 2000). A decrease in DNA content due to freezing is reported for soil samples, where mainly the archaeal communities are affected

(Pesaro et al. 2003). Freezing and thawing had most impact on ATP concentrations which were used to estimate the number of active cells. We speculate that inactive or dormant cells, which may constitute a considerable portion of natural groundwater microbial communities (Alfreider et al. 1997, Kieft & Phelps 1997), are less sensitive to freezing. Likely the number of inactive cells is comparable in high activity and low activity samples of a given site, which may explain the non-proportional loss of cells by freezing. The robust, less active cells seem to persist best and are not sensitive to repeated freezing and thawing treatments. If mainly active cells are lost, biomass and especially community analysis are biased after freezing, as these members of the communities are of most interest. Furthermore, low nucleic acid bacteria, which have been shown to be more resistant to oxidative disinfection (Ramseier et al. 2011) may also be more resistant to freezing. Unfortunately, fixation is possible only for direct cell counting but not for indirect methods such as ATP and 16S rDNA copies where freezing is still the only possible method to preserve the status in a sample after sampling.

Finally, the effects of freezing have been tested with sediment samples collected along a vertical gradient in an aquifer contaminated with aromatic hydrocarbons. Here, the total numbers of bacterial cells have been determined with fresh glutardialdehyde fixed sediments as well as with aliquots after freezing at -20°C and thawing at room temperature. The higher the absolute cell number was in the freshly fixed samples the higher was the relative loss of cells due to freezing, with exception of the uppermost sampling location (Fig. 4.6). Cell loss averaged 32%, but ranged from 20 to 90%. Again, there was no significant difference between cell counts in samples frozen once or repeatedly frozen and thawed for two or three times (data not shown, $p < 0.05$).

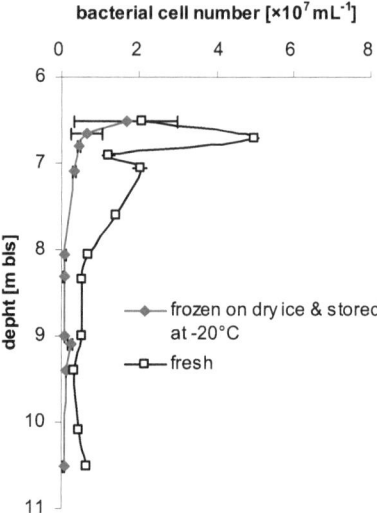

Figure 4.6: Vertical distribution of total bacteria as determined in a sediment core from a tar oil contaminated sandy aquifer from freshly processed samples and parallel samples which have been frozen on dry ice and stored at -20°C. Values are triplicates ± SD.

4.4. Conclusion

The paper introduces a time-saving, robust and reliable quantification of sediment bacteria by means of flow cytometry. Moreover, various technical aspects ranging from the dislodgement of attached cells, using an appropriate fluorescent dye as well as sample purification with DGC are discussed, which will allow a systematic adaptation of our approach to other types of sediment. Furthermore, we could demonstrated that freezing and thawing of sediment samples results in substantial loss of total and active cells, as well as the number of 16S rDNA copies. There is evidence that the cells lost come from the formerly active fraction of the samples. Repeated freezing does not lead to a further significant decrease in cell numbers, hinting at cells resistant to freezing and thawing. Taking into account that freezing is the standard preservation treatment prior to molecular analysis, the loss of DNA is crucial, especially if it represents the originally most active members of the consortium.

4.5. References

Alfreider A, Krössbacher M, Psenner R (1997) Groundwater samples do not reflect bacterial densities and activity in surface systems. Water Research 31:882-840

Amalfitano S, Fazi S (2008) Recovery and quantification of bacterial cells associated with streambed sediments. Journal of Microbiological Methods 75:237-243

Anneser B, Einsiedl F, Meckenstock RU, Richters L, Wisotzky F, Griebler C (2008) High-resolution monitoring of biochemical gradients in a tar oil-contaminated aquifer. Applied Geochemistry 23:1715-1730

Anneser B, Pilloni G, Bayer A, Lueders T, Griebler C (2010) High Resolution Analysis of Contaminant Aquifer Sediments and Groundwater – What Can be learned in Terms of Natural Attenuation? Geomicrobiology Journal 27:130-142

Bach H-J, Tomanova J, Schloter M, Munche JC (2002) Enumeration of total bacteria and bacteria with genes for proteolytic activity in pure cultures and in environmental samples by quantitative PCR mediated amplification. Journal of Microbiological Methods 49:235-245

Bachoon DS, Chen F, Hodson RE (2001) RNA recovery and detection of mRNA by RT-PCR from preserved prokaryotic samples. FEMS Microbiology Letters 201:127-132

Bakken LR, Lindahl V (1995) Recovery of bacterial cells from soil. In: van Elsas JD, Trevors JT (eds) Nucleic Acids in the Environment: Methods and Applications. Springer Verlag, Heidelberg, p 9-27

Bast E (2001) Mikrobiologische Methoden – Eine Einführung in grundlegende Arbeitstechniken, 2 edn. Spektrum Akademischer Verlag, Heidelberg Berlin

Berney M, Vital M, Hülshoff I, Weilenmann H-U, Egli T, Hammes F (2008) Rapid, cultivation-independent assessment of microbial viability in drinking water. Water Research 42:4010-4018

Biddle JF, Fitz-Gibbon S, Schuster SC, Brenchley JE, House CH (2008) Metagenomic signatures of the Peru Margin subseafloor biosphere show a genetically distinct environment. PNAS 105:10583-10588

Böckelmann U, Szewzyk U, Grohmann E (2003) A new enzymatic method for the detachment of particle associated soil bacteria. Journal of Microbiological Methods 55:201-211

Boenigk J (2004) A disintegration method for direct counting of bacteria in clay-dominated sediments: dissolving silicates and subsequent fluorescent staining of bacteria. Journal of Microbiological Methods 56:151-159

Buesing N, Gessner MO (2002) Comparison of detachment procedures for direct counts of bacteria associated with sediment particles, plant litter and epiphyctic biofilms. Aquatic Microbial Ecology 27:29-36

Chang BV, Lu YS, Yuan SY, Tsao TM, Wang MK (2009) Biodegradation of phtalate esters in compost-amended soil. Chemosphere 74:873-877

Courtois S, Frostegard A, Göransson P, Depret G, Jeannin P, Simonet P (2001) Quantification of bacterial subgroups in soil: comparison of DNA extracted directly from soil or from cells previously released by density gradient centrifugation. Environmental Microbiology 3:431-439

Danielopol DL, Griebler C, Gunatilaka A, Notenboom J (2003) Present state and future prospects for groundwater ecosystems. Environmental Conservation 30:104-130

Davey HM, Kell DB (1996) Flow cytometry and cell sorting of heterogeneous microbial populations: The importance of single-cell analyses. Microbiological Reviews 60:641-696

Dayley RJ, Hobbie JE (1975) Direct counts of aquatic bacteria by a modified epifluorescence technique. Limnology and Oceanography 20:875-882

Dodd CER, Richards PJ, Aldsworth TG (2007) Suicide through stress: A bacterial response to sublethal injury in the food environment. International Journal of Food Microbiology 120:46-50

Dodd CER, Sharman RL, Bloomfield SF, Booth IR, Stewart GSAB (1997) Inimical processes: Bacterial self-destruction and sub-lethal injury. Trends in Food Science & Technology 8:238-241

Duhamel S, Jacquet S (2006) Flow cytometric analysis of bacteria- and virus-like particles in lake sediments. Journal of Microbiological Methods 64:361-332

Ellery WN, Schleyer MH (1984) Comparison of homogenization and ultrasonication as techniques in extracting attached sediment bacteria. Marine Ecology Progress Series 15:247-250

Ellison A, Perry SF, Stewart GSAB (1991) Bioluminescence as a real-time monitor of injury and recovery in *Salmonella typhimurium*. International Journal of Food Microbiology 12:323-332

Epstein SS, Rossel J (1995) Enumeration of sandy sediment bacteria: search for optimal protocol. Marine Ecology Progress Series 117:289-298

Fogel GB, Collins CR, Li J, Brunk CF (1999) Prokaryotic Genome Size and SSU rDNA Copy Number: Estimation of Microbial Relative Abundance from a Mixed Population. Microbial Ecology 38:93-113

Gasol JM, Zweifel UL, Peters F, Fuhrmann JA, Hagström A (1999) Significance of Size and Nucleic Acid Content Heterogeneity as Measured by Flow Cytometry in Natural Planktonic Bacteria. Applied and Environmental Microbiology 65:4475-4483

Gough HL, Stahl DA (2003) Optimization of direct cell counting in sediment. Journal of Microbiological Methods 52:39-46

Griebler C (2001) Microbial ecology of subsurface ecosystems. In: Griebler C, Danielopol DL, Gibert J, Nachtnebel HP, Notenboom J (eds) Groundwater ecology – a tool for management of water resources. Office of Official Publications of the European Communities, Luxembourg, p 81-108

Griebler C, Lueders T (2009) Microbial biodiversity in groundwater ecosystems. Freshwater Biology 54:649-677

Griebler C, Mindl B, Slezak D (2001) Combining DAPI and SYBR Green II for the Enumeration of Total Bacterial Numbers in Aquatic Sediments. International Review of Hydrobiology 86:453-465

Griebler C, Mindl B, Slezak D, Geiger-Kaiser M (2002) Distribution patterns of attached and suspended bacteria in pristine and contaminated shallow aquifers studied with an in situ sediment exposure microcosm. Aquatic Microbial Ecology 28:117-129

Griebler C, Wirth N, Mindl B (1997) Ein neuer Ansatz zur Bestimmung der bakteriellen Abundanz in Bach- und Flusssedimenten. Methodenvergleich und Effizienzbestimmung. 32. Konferenz der Internationalen Arbeitsgemeinschaft Donauforschung der SIL, Wien, p 127-130

Gruden C, Skerlos S, Adriaens P (2004) Flow cytometry for microbial sensing in environmental sustainability applications: current status and future prospects. FEMS Microbiology Ecology 49:37-49

Gruden C, Khijniak A, Adriaens P (2003) Activity assessment of microorganisms eluted from sediments using 5-cyano-2,3-ditolyl tetrazolium chloride: a quantitative comparison of flow cytometry to epifluorescent microscopy. Journal of Microbiological Methods 55:865-874

Günther S, Hübschmann T, Rudolf M, Eschenhagen M, Röske I, Harms H, Müller S (2008) Fixation procedures for flow cytometric analysis of environmental bacteria. Journal of Microbiological Methods 75:127-134

Hammes F, Berney M, Wang Y, Vital M, Köster O, Egli T (2008) Flow-cytometric total bacterial cell counts as a descriptive microbial parameter for drinking water treatment processes. Water Research 42:269-277

Hammes F, Egli T (2005) New Method for Assimilable Organic Carbon Determination Using Flow-Cytometric Enumeration and a Natural Microbial Consortium as Inoculum. Environmental Science & Technology 39:3289-3294

Hammes F, Goldschmidt F, Vital M, Wang Y, Egli T (2010) Measurement and interpretation of microbial adenosine tri-phosphate (ATP) in aquatic environments. Water Research 44:3915-3923

Helton RR, Liu L, Wommack KE (2006) Assessment of Factors Influencing Direct Enumeration of Viruses within Estuarine Sediments. Applied and Environmental Microbiology 72:4767-4774

Joachimsthal EL, Ivanov V, Tay JH, Tay STL (2003) Flow cytometry and conventional enumeration of microorganisms in ships' ballast water and marine samples. Marine Pollution Bulletin 46:308-313

Jones SG (1974) Some observations on direct counts of freshwater bacteria obtained with a fluorescence microscope. Limnology and Oceanography 19:540-543

Kallmeyer J, Smith DS, Spivack AJ, D'Hondt S (2008) New cell extraction procedure applied to deep subsurface sediments. Limnology and Oceanography: Methods 6:236-245

Kieft TL, Phelps TJ (1997) Life in the slow line: Activities of microorganisms in the subsurface. In: Amy PS, Haldeman DL (eds) The Microbiology of the Terrestrial Subsurface, Boca Raton, p 137-163

Kirchman D, Sigda J, Kapuscinski R, Mitchell R (1982) Statistical Analysis of the Direct Count Method for Enumerating Bacteria. Applied and Environmental Microbiology 44:376-382

Klappenbach JA, Dunbar J, M., Schmidt TM (2000) rRNA Operon Copy Number Reflects Ecological Strategies of Bacteria. Applied and Environmental Microbiology 66:1328-1333

Klauth P, Wilhelm R, Klumpp E, Poschen L, Groeneweg J (2004) Enumeration of soil bacteria with the green fluorescent nucleic acid dye Sytox green in the presence of soil particles. Journal of Microbiological Methods 59:189-198

Kuwae T, Hosokawa Y (1999) Determination of abundance and biovolume of bacteria in sediments by dual staining with 4',6-diamidino-2-phenylindole and acridine orange: Relationship to dispersion treatment and sediment characteristics. Applied and Environmental Microbiology 65:3407-3412

Lebaron P, Parthuisot N, Catala P (1998) Comparison of blue nucleic acid dyes for flow cytometric enumeration of bacteria in aquatic systems. Applied and Environmental Microbiology 64:1725-1730

Lindahl V (1996) Improved soil dispersion procedures for total bacterial counts, extraction of indigenous bacteria and cell survival. Journal of Microbiological Methods 25:279-286

Lindahl V, Bakken LR (1995) Evaluation of methods for extraction of bacteria from soil. FEMS Microbiology Ecology 16:135-142

Lucchesi P, Santangelo G (1997) The intersitial ciliate microcommunity of a Mediterranean sandy shore under differing hydrodynamic disturbances. Italian Journal of Zoology 64:253-259

Lueders T, Manefield M, Friedrich MW (2004) Enhanced sensitivity of DNA- and rRNA-based stable isotope probing by fractionation and quantitative analysis of isopycnic centrifugation gradients. Environmental Microbiology 6:73-78

Lunau M, Lemke A, Walther K, Martens-Habbena W, Simon M (2005) An improved method for counting bacteria from sediments and turbid environments by epifluorescence microscopy. Environmental Microbiology 7:961-968

Madsen EL (2000) Nucleic-acid characterization of the identity and activity of subsurface microorganisms. Hydrogeology Journal 8:112-125

Martens-Habbena W, Sass H (2006) Sensitive Determination of Microbial Growth by Nucleic Acid Staining in Aqueous Suspension. Applied and Environmental Microbiology 72:87-95

Mayr C, Winding A, Hendriksen NB (1999) Community level physiological profile of soil bacteria unaffected by extraction method. Journal of Microbiological Methods 36:29-33

Mermillod-Blondin F, Fauvet G, Chalamet A, Creuzé des Châtelliers M (2001) A Comparison of Two Ultrasonic Methods for Detaching Biofilms from Natural Substrata. International Review of Hydrobiology 86:349-360

Meyer-Reil L-A (1993) Mikrobielle Besiedlung und Produktion. In: Meyer-Reil L-A, Köster M (eds) Mikrobiologie des Meeresbodens. Gustav Fischer, Jena, Stuttgart, New York, p 38-75

Muyzer G, Teske A, Wirsen CO, Jannasch HW (1995) Phylogenetic relationships of *Thiomicrospira* species and their identification in deep-sea hydrothermal vent samples by denaturating gradient gel electrophoresis of 16S rDNA fragments. Arch Microbiol 146:164-172

Nunan N, Wu KJ, Young IM, Crawford JW, Ritz K (2003) Spatial distribution of bacterial communities and their relationships with the micro-architecture of soil. FEMS Microbiology Ecology 44:203-215

Pesaro M, Widmer F, Nicollier G, Zeyer J (2003) Effects of freeze-thaw stress during soil storage on microbial communities and methidathion degradation. Soil Biology & Biochemistry 35:1049-1061

Porter KG, Feig YS (1980) The use of DAPI for identifying and counting aquatic microflora. Limnology and Oceanography 25:943-948

Priemé A, Sitaula JIB, Klemedtsson AK, Bakken LR (1996) Extraction of methane-oxidizing bacteria from soil particles. FEMS Microbiology Ecology 21:59-86

Rabus R, Widdel F (1995) Anaerobic degradation of ethylbenzene and other aromatic hydrocarbons by new denitrifying bacteria. Archives of Microbiology 163:96-103

Ramseier MK, von Gunten U, Freihofer P, Hammes F (2011) Kinetics of membrane damage to high (HNA) and low (LNA) nucleic acid bacterial cluster in drinking water by ozone, chlorine, chlorine dioxide, monochloramine, ferrate (VI), and permanganate. Water Research 45:1490-1500

Schallenberg M, Kalff J, Rasmussen JB (1989) Solutions to problems in enumerating sediment bacteria by direct counts. Applied and Environmental Microbiology 55:1214-1219

Skogland T, Lomeland S, Goksoyr J (1988) Respiratory burst after freezing and thawing of soil: experiments with soil bacteria. Soil Biology & Biochemistry 20:851-856

Velji IM, Albright LJ (1993) Improved sample preparation for enumeration of aggregated aquatic substrate bacteria. In: Kemp PF, Sherr EB, Cole JJ (eds) Handbook of methods in aquatic microbial ecology. Lewis Publisher, Boca Raton, p 139-142

Velten S, Hammes F, Boller M, Egli T (2007) Rapid and direct estimation of active biomass on granular activated carbon through adenosine tri-phosphate (ATP) determination. Water Research 41:1973-1983

Vives-Rego J, Lebaron P, Nebe-von Caron G (2000) Current and future applications of flow cytometry in aquatic microbiology. FEMS Microbiology Reviews 24:429-448

Wang Y, Hammes F, Boon N, Chami M, Egli T (2009) Isolation and characterization of low nucleic acid (LNA)-content bacteria. ISME 3:889-902

Weinbauer MG, Beckmann C, Hofle MG (1998) Utility of green fluorescent nucleic acid dyes and aluminum oxide membrane filters for rapid epifluorescence enumeration of soil and sediment bacteria. Applied and Environmental Microbiology 64:5000-5003

Whitman BW, Coleman DC, Wiebe W (1998) Prokaryotes: The unseen majority. PNAS 95:6578-6583

Winderl C, Anneser B, Griebler C, Meckenstock RU, Lueders T (2008) Depth-resolved quantification of anaerobic toluene degraders and aquifer microbial community patterns in distinct redox zones of a tar oil contaminant plume. Applied and Environmental Microbiology 74:792-801

Witzingerode F, Göbel UB, Stackebrandt E (1997) Determination of microbial diversity in environmental samples: pitfalls of PCR-based rRNA analysis. FEMS Microbiology Ecology 21:213-229

Ziglio G, Andreottola G, Barbesti S, Boschetti G, Bruni L, Foladori P, Villa R (2002) Assessment of activated sludge viability with flow cytometry. Water Research 36:460-468

5. General discussion and conclusion

Biodegradation of organic dissolved contaminants in porous aquifers may be limited by a number of factors such as the availability of appropriate electron acceptors and nutrients, kinetic limitations, or simply by the low number of degraders (e.g. Lerner et al. 2000). Transverse dispersion is known to be the main factor controlling the mixing of electron donors (contaminants) and electron acceptors (e.g. oxygen, sulfate) which is essential for microbial biodegradation. In consequence, highest biodegradation activity is located at transition zones between contaminated and non-contaminated areas such as the fringes of contaminant plumes (e.g. Thornton et al. 2001, Bauer et al. 2008). Hence, biodegradation may be restricted to a small area in the centimeter to decimeter scale and pronounced bioactivities are indicated by steep physical-chemical and microbiological gradients (Anneser et al. 2008, Bauer et al. 2008). The detection of these gradients is only possible with sampling at a very high vertical resolution. Conventional multi-level wells (C-MLW) have a spatial resolution (e.g. 0.5 to 1 m at the site in Düsseldorf) which is far to low detecting the small-scale changes observed for contaminants and redox-sensitive parameters. Thus, small spatial variations are often overseen. Furthermore, sampling C-MLWs is affected by disturbing the gradients by mixing water of different depths during pumping. The special high-resolution multi-level well (HR-MLW) installed at the investigated site in Düsseldorf-Flingern in June 2005, overcomes these difficulties.

Additional to sampling of groundwater sediments were collected and also analyzed in a high vertical resolution to compare the data gained from the water samples. One aim was to know, if sampling and analyzing of groundwater only is sufficient to uncover the most important NA processes in the subsurface. In chapter 2 the individual importance as well as the differences of information gained from groundwater and sediment samples are discussed. It is demonstrated that the situation in the groundwater reflects the conditions in the aquifer only to a certain degree. With sampling groundwater it was possible to identify and localize redox conditions, but merely the current conditions are reflected. Data gained from the groundwater can be rather seen as a "snapshot", whereas the sediment stores information from the past in its "long-term memory". Hence, the history of the underground is better depicted by sampling sediment. Without doubt, sediment samples need to be considered when dealing with high molecular compounds such as PAHs, which tend to sorb to particle surfaces to a high degree. Moreover, analysis of sediments is essential from a microbiological perspective, *i.e.* for the quantification of bacterial cells, analysis of community composition and microbial activity. To get quantitatively reliable data on the distribution of contaminants and microorganisms, sampling and comparing groundwater as well as sediment samples remain indispensable.

So far, porous groundwater systems have been considered to be relatively stable in their environmental conditions. Indeed, it is known that karst systems are hydraulically dynamic and it has been shown that precipitation as well as flooding events directly influence porous aquifers. With respect to biological patterns it is rather unclear in which timescales these abiotic dynamics may affect NA processes in aquifers. In detail, it is not known how fast the associated degrader communities can adapt to quickly changing conditions. Assuming pronounced hydrological dynamics, two contrary aspects are possible. (1) On the one hand transient conditions may enhance the NA capacity by improved distribution of the contaminant as well as electron acceptors over a larger reactive volume. Whereas (2) on the other hand transient conditions may impair the NA capacity, posing temporally unfavorable conditions to the highly specialized sessile degrader communities. The consecutive lack of the electron acceptor or the electron donor (contaminant) slows down the biodegradation activity.

Long-term changes (time range of months) of physical-chemical and biological gradients in the groundwater have already been shown in a preceding study (Anneser 2008). Now, the focus were dynamics of physical-chemical and microbial gradients vertically across a BTEX plume with inherent sulfide allocation coupled to sulfate reduction as the major redox process linked to NA. Indeed, changes were detectable after only two weeks at the test site (chapter 3). A groundwater table drop of only 3 cm in this short time span implicated a downward shift of the lower plume fringe of 10 cm and consequently a concomitant vertical expansion of the BTEX-plume. The major sulfide peaks were always located at the lower plume fringe and followed the plume. This temporally highly resolved data indicate either a fast adaption of degradation and establishment of NA activity or hint at a broad distribution of degrader populations which switch on and off their activity along with environmental changes. Considering the relatively slow doubling times of groundwater bacteria, which generally range from months to years (Mailloux & Fuller 2003, Griebler & Lueders 2009), it is improbable that degraders establish new populations in only two weeks. Moreover, as the bulk of bacteria are attached to sediment particles it is unlikely that they readily move from one location to another. Therefore, it is suggested that these observations can not be assigned to a frequent indigenous augmentation of microbial biomass, rather than to a change in microbial degradation activity.

During the present work stronger variations of the maximum BTEX concentrations than ever detected could be observed and are highlighted in chapter 3. In 2008 the BTEX concentration at our HR-MLW reached the highest value measured since the installation of the well. Surprisingly, concentration changes were found only with toluene and xylene, but not with e.g. benzene and naphthalene. Based on this observation two different theories were established. The first guess comprises a collapse of the toluene biodegradation caused by a collapse of the specific degrader populations, probably but not necessary because of higher toluene concentrations that reached toxic values. Alternatively, the contaminant composition of the plume changed to higher toluene and xylene concentrations, without significant changes in the total biodegradation activity. Such a dynamic could have been attributed to the groundwater table elevation and immersion of parts of the source, which have been in the unsaturated zone before. There is also the possibility of an individual second source in the formerly unsaturated zone. The contribution of contaminants from a new source is supported by the change in $^{13}C/^{12}C$ toluene isotope signatures. Different contaminant sources may have different isotopic signatures and also sources of the same age and production line may have a different signature being in the unsaturated zone than in the saturated zone for certain periods of time (Hunkeler et al. 2004). The collapse of toluene and xylene degradation was indicated by received support from bacterial community analysis. Giovanni Pilloni, a PhD student working on water and sediment samples from selected sampling occasions, could show a break down of the major toluene degrading population, *i.e.* representatives of the *Desulfobulbaceae*, along with a strong reduction of the abundance of the *bssA*-gene copies compared to total 16S rDNA-gene copies (Pilloni 2011). The *bssA*-gene is encoding a subunit of the key enzyme for anaerobic toluene degradation (Winderl et al. 2008). In detail, a sulfate reducing community degrading toluene was dominated by *Desulfobulbaceae* (belonging to the *Deltaproteobacteria*) and established at the lower plume fringe, but experienced a dramatic collapse of the *Desulfobulbaceae* population in 2008. Subsequently, a distinct but functionally redundant population of degraders within the *Clostridia*, i.e. *Peptococcaceae* established in 2009 (Pilloni 2011).

Overall evidence of biodegradation of toluene and sulfate reduction was repeatedly proven by compound specific isotope analysis (CSIA) of $^{13}C/^{12}C$ in toluene as well as of $^{18}O/^{16}O$ and $^{34}S/^{32}S$ in sulfate. The compound specific stable isotope fractionation can be clearly assigned to biological processes. Thus, the CSIA approach was used for an approximate estimation of the amount of biodegradation (Rayleigh 1896, Richnow et al. 2003). With this approach, also the theoretical concentration of toluene neglecting biodegradation was calculated. These calculated initial toluene concentrations revealed smaller than the initial calculated toluene concentration when assuming the difference of background sulfate and residual sulfate was used entirely for just toluene degradation (Anneser et al. 2008). This finding is not surprising, as there are a number of other contaminants which are likely linked to sulfate reduction. Only in September 2008, both calculation approaches revealed the same initial toluene concentrations (Appendix Fig. A.2). What means that the sulfate was either only used for degradation of toluene, or, what is more likely, mainly used for the degradation of other contaminants. It is worth to mention that CSIA of sulfate shows an ongoing sulfate reduction.

Besides the high toluene and xylene concentrations in September 2008 and the changes in the ratio between toluene and xylene and the other contaminants, i.e. ethylbenzene and naphthalene, we collected independent evidence of a collapse in microbial toluene degradation by CSIA of toluene. In February 2006 and 2007 a steep and small scale gradient of the toluene $\delta^{13}C$ values was found clearly pointing pronounced toluene biodegradation activities in these zones, from -24.5‰ in the plume core to -23‰ to -20.5‰ at the upper and lower plume fringes. This destructive pattern was lost in September 2008. Although the overall toluene isotope signature showed enriched values of about -23‰ to -22.5‰ no more pronounced gradients could be found. At the moment, the most likely explanation of the enriched isotope values is a changed source signature related to the contribution of a new individual source or part of the one source formerly located in the unsaturated zone. Impressively, the isotopic gradients partly reestablished in 2009, mainly at the upper plume fringe, supporting the data from community analysis, namely the establishment of a new degrader population and toluene degrading activity.

The pattern of high BTEX concentrations in 2008 and the very low concentrations in 2009 could be further confirmed by data from a neighboring C-MLW sampled and analyzed by the Stadtwerke Düsseldorf.

Holling called the capacity of a natural system to cope with disturbances and restoring its functional state the "ecosystem resilience" (Holling 1973). When observing the investigated site over a very long period (years), it seems that functional redundancy within the microbial community can insure the NA capacity in this aquifer section against the disturbances. The degrader community dominated by *Desulfobulbaceae* was replaced by a new population dominated by *Peptococcaceae* after about two years, and biodegradation of toluene went on again. Whereas when this site is regarded over a shorter period (months), the disturbances mainly related to hydrological dynamics have a strong impact on biodegradation capacity. This outcome of the study points at the above mentioned aspect two, that "the transient hydraulic conditions temporally impair the efficient contaminant degradation by specialized degrader populations" and lead to an overall decrease in biodegradation rather than an increase. Moreover, active degraders have to proliferate and the slow doubling times of microorganisms in these kind of environment has to be taken into consideration. On the other hand resilience of NA processes to small hydrodynamics is obvious looking at the short-term dynamics in 2009. The vertical expansion of the BTEX plume changed measurably within only two weeks, but the sulfide peaks, as indications of biodegradation, moved always together with the lower plume fringe. A broad distribution of degraders adapted or resistant to redox changes switch on and off biodegradation, supposable neglecting moving or proliferating in this short period of time.

The final part of this discussion is dedicated to the quantification of bacteria in sediment samples, which constitutes an important aspect of my investigations. Analyzing sediment samples is always a great challenge, because of the poor accessibility of the subsurface. Sampling groundwater repeatedly from an established well is rather easy, whereas repeatedly collecting sediment samples is related to serious financial and technical efforts. For example, it is impossible to sample at exact the same position twice. Due to the drilling the soil is partly removed and the drilling hole is either filled by new additional material or spontaneously filled by surrounding material collapsing. A further difficulty is the analysis of sediment samples. In detail, to enumerate bacteria quantitative detaching and separating the bacterial cells from the inorganic particles is crucial. During my studies a relatively fast and reliable method was developed for reliable enumeration of bacteria which is described in chapter 4. As a first step, the cells are removed from sediments without causing disruption. Subsequently, cells are separated from inorganic particles via density gradient centrifugation followed by quantification by means of flow cytometry. This quantification approach was successfully applied to sediment samples of the Düsseldorf-Flingern aquifer, but can easily be adapted to different other kinds of sediment. Flow cytometry is a regularly used tool assessing the total cell number, cell size and viability in aquatic samples. Our approach will allow the determination of these parameters also with sediment samples in the future.

Moreover, the influence of the storage of samples on the total cell number, ATP concentration as well as the total number of 16S rDNA gene copies was investigated. It could be shown for the first time that freezing reduces not just cell numbers but also gene copies. Until now, freezing is the commonly used "fixation" method prior to DNA and RNA extraction followed by further molecular analysis. We thus should have in mind that freezing can not only decrease the cell number but may also change the bacterial community composition. Further research need to be addressed which part of microbial communities is mainly affected.

The investigated field site in Düsseldorf-Flingern is unique with its data set collected over almost five years taking into account the extraordinary high spatial and temporal resolution. An important part of work remaining is to translate the generic findings into application to enhance NA at sites contaminated with hydrocarbons. Düsseldorf-Flingern can be seen as a model field site. Partly the data have been already implemented and tested by modelers (Prommer et al. 2009). Improvement of current numerical models and model runs with the data may help to hint at further processes important for NA *in situ*. There are still some interesting and important questions which remain. First of all, the data indicate further factors limiting biodegradation besides transverse and longitudinal dispersion (mixing). We need to understand better growth patterns and dynamics of degraders. It also needs to be tested if the gained observations and results can be transferred to similar polluted sites and if the dynamics observed at this test site are also detectable at other polluted sites. It could be demonstrated that hydraulic dynamics of the groundwater table can have a strong influence on the distribution of redox gradients as well as on biodegradation, have an inhibiting effect on microbial contaminant degradation and, hence, influence the overall concentration and distribution of contaminants. Individual populations of highly specialized degraders are found to be very vulnerable to changing environmental conditions, but carrying a certain adaption possibility. When regarding events over a very long period of time, also because the long doubling times of inherent soil microorganisms, the ecosystem with its communities can ensure its natural attenuation potential.

References

Anneser B (2008) Spatial and temporal dynamics of biochemical gradients in a tar oil-contaminated porous aquifer – biodegradation processes revealed by high-resolution measurements. Eberhard-Karls-Universität, Tübingen

Anneser B, Einsiedl F, Meckenstock RU, Richters L, Wisotzky F, Griebler C (2008) High-resolution monitoring of biochemical gradients in a tar oil-contaminated aquifer. Applied Geochemistry 23:1715-1730

Bauer RD, Maloszewski P, Zhang Y, Meckenstock RU, Griebler C (2008) Mixing-controlled biodegradation in a toluene plume – Results from two-dimensional laboratory experiments. Journal of Contaminant Hydrology 96:150-168

Griebler C, Lueders T (2009) Microbial biodiversity in groundwater ecosystems. Freshwater Biology 54:649-677

Holling CS (1973) Resilience and Stability of Ecological Systems. Annual Review of Ecology and Systematics 4:1-23

Hunkeler D, Chollet N, Pittet X, Aravena R, Cherry JA, Parker BL (2004) Effect of source variability and transport processes on carbon isotope ratios of TCE and PCE in two sandy aquifers. Journal of Contaminant Hydrology 74:265-282

Lerner DN, Thornton SF, Spence MJ, Banwart SA, Bottrell SH, Higgo JJ, Mallison HEH, Pickup RW, Williams GM (2000) Ineffective Natural Attenuation of Degradable Organic Compounds in a Phenol-Contaminated Aquifer. Groundwater 38:922-928

Mailloux BJ, Fuller ME (2003) Determination of In Situ Bacterial Growth Rates in Aquifers and Aquifer Sediments. Applied and Environmental Microbiology 69:3798-3808

Pilloni G (2011) Distribution and dynamics of contaminant degraders and microbial communitites in stationary and non-stationary contaminant plumes. Technische Universität, München

Prommer H, Anneser B, Rolle M, Einsiedl F, Griebler C (2009) Biogeochemical and Isotopic Gradients in a BTEX/PAH Contaminant Plume: Model-Based Interpretation of a High-Resolution Field Data Set. Environmental Science & Technology 43:8206-8212

Rayleigh L (1896) Theoretical considerations respecting the separation of gases by diffusion and similar processes. Philosophical Magazine Series 5 42:259, 493-498

Richnow HH, Annweiler E, Michaelis W, Meckenstock RU (2003) Microbial in situ degradation of aromatic hydrocarbons in a contaminated aquifer monitored by carbon isotope fractionation. Journal of Contaminant Hydrology 65:101-120

Thornton SF, Quigley S, Spence MJ, Banwart SA, Bottrell S, Lerner DN (2001) Processes controlling the distribution and natural attenuation of dissolved phenolic compounds in a deep sandstone aquifer. Journal of Contaminant Hydrology 53:233-267

Winderl C, Anneser B, Griebler C, Meckenstock RU, Lueders T (2008) Depth-resolved quantification of anaerobic toluene degraders and aquifer microbial community patterns in distinct redox zones of a tar oil contaminant plume. Applied and Environmental Microbiology 74:792-801

Appendix

Table A.1: Concentrations of BTEX, toluene, stable isotope values of toluene ($^{13}C/^{12}C$), PAHs, redox, sulfide, sulfate, stable isotope values of sulfate ($^{34}S/^{32}S$ and $^{18}O/^{16}O$) and bacterial biomass in groundwater sampled with the high-resolution multi-level well in September 2008. av. = average, s.d. = standard deviation, b.d. = below detection limit.

Depth [m bls]	BTEX [mg L^{-1}] av.	s.d.	Toluene [mg L^{-1}] av.	s.d.	$\delta^{13}C$ [‰]	PAHs [mg L^{-1}] av.	s.d.	Redox [mV]
6.385	32.77	1.48	21.982	0.776	-23.07	2.58	2.13	-81
6.410	40.30	2.85	28.358	2.639	-22.38	4.54	0.18	-71
6.435	54.26	5.48	41.974	3.936	-22.68	2.90	0.40	-79
6.460	72.03	13.95	56.017	9.857	-22.35	4.45	0	-90
6.510	121.72	14.79	97.267	12.656	-22.62		0.16	-120
6.535	116.61	35.69	91.760	25.098	-22.16		0	-145
6.585	72.23	12.30	59.571	9.956	-22.65	4.19	0.13	-189
6.610	70.75	5.76	59.218	4.138	-22.69	3.22	0.06	-176
6.635	75.42	20.42	62.251	16.680	-22.39	4.12	0.20	-167
6.665	75.77	7.12	63.160	4.540	-22.14	5.03	0.22	-176
6.695	37.61	6.78	29.990	4.228	-22.44	4.58	0	-187
6.745	32.65	2.58	27.652	1.746	-22.85	2.55	2.88	-199
6.805	11.61	4.66	9.981	3.709	-23.03	2.64	2.72	-172
6.855	6.99	0.25	5.308	0.094	-23.15	1.71	1.52	-216
6.905	2.15	0.13	1.559	0.123	-22.24	0.94	0.19	-229
6.955	0.47	0.23	0.225	0.225	-23.71	1.86	1.08	-215
7.005	0.21	0.06	0.176	0.040	b.d.	1.69	0	-209
7.055	0.22	0.01	0.186	0.001	b.d.	1.87	0.13	-136
7.105	0.13	0.04	0.103	0.033	b.d.	1.90	0.18	-85
7.155	0.15	0.05	0.127	0.044	b.d.	1.98	0.09	-61
7.205	0.12	0.06	0.105	0.058	b.d.	2.44	0.32	-75
7.255	0.18	0.02	0.170	0.015	b.d.	2.37	0.12	-80
7.305	0.19	0.04	0.152	0.042	b.d.	2.25	0	-81
7.405	0.20	0.02	0.172	0.017	b.d.	2.48	0.14	-87
7.505	0.27	0.11	0.225	0.090	b.d.	2.31	0.14	-67
7.605	0.29	0.09	0.248	0.083	b.d.	2.21	0.10	-80
8.045	0.26	0.10	0.206	0.074	b.d.	0.80	0.30	-106
9.045	0.24	0.14	0.227	0.136	b.d.	0.78	0.36	-60
10.195	0.33	0.11	0.328	0.100	b.d.	1.12	0.06	-40
11.185	0.25	0.10	0.241	0.097	b.d.	0.56	0.10	-35

Table A.1 continued

Depth [m bls]	H_2S [mg L^{-1}] av.	s.d.	SO_4^{2-} [mg L^{-1}] av.	s.d.	$\delta^{34}S$ [‰]	$\delta^{18}O$ [‰]	Cells [x10^6 mL^{-1}] av.	s.d.
6.385	b.d.		0.65	0.05	b.d.		7.41	9.12
6.410	0.032	0.032	0.50	0.00	b.d.		8.32	6.36
6.435	0.064	0	2.20	0.10	b.d.		5.53	3.04
6.460	0.191	0	3.55	0.15	b.d.		3.30	2.12
6.510	0.127	0	4.85	4.05	b.d.		3.31	2.33
6.535	1.462	0	7.35	0.15	b.d.		1.26	4.80
6.585	5.053	0.095	15.55	0.35	35.21		3.09	5.34
6.610	4.703	0.381	24.90	0.10	27.76	11.96	2.13	1.28
6.635	4.068	0	32.10	0.20	29.68		1.56	6.51
6.665	6.165	0.127	38.65	0.15	28.87	11.83	2.26	1.80
6.695	5.117	0.095	42.45	0.45	36.98	15.19	3.57	2.92
6.745	6.515	0.350	66.80	0.30	33.66	13.85	1.79	1.40
6.805	8.199	0.127	103.25	0.05			6.18	4.47
6.855	8.453	0.064	128.80	0.00	32.18		4.26	3.76
6.905	10.169	0.064	145.25	0.85	28.95	13.24	5.70	5.78
6.955	7.214	0.159	157.60	0.60	26.43		6.76	7.14
7.005	2.288	0.191	166.70	1.80	24.86	12.21	9.96	9.40
7.055	1.049	0.032	173.40	2.00	24.18		1.24	9.69
7.105	0.350	0.032	182.30	2.10	23.20	12.51	1.40	7.48
7.155	0.318	0	189.50	1.60	22.34		2.09	1.89
7.205	0.381	0	194.75	1.35	21.41	11.90	1.61	7.72
7.255	0.191	0	200.70	0.60	20.79		2.16	9.82
7.305	0.254	0	199.35	0.55	20.76	12.36	1.60	6.82
7.405	0.222	0.159	190.70	2.20	22.12	12.36	1.44	1.33
7.505	0.095	0.032	183.85	0.75	22.78		1.58	1.62
7.605	0.095	0.032	186.90	0.20	21.52	12.27	2.58	8.34
8.045	3.337	0.095	176.00	0.40	23.02		1.03	5.55
9.045	0.032	0.032	296.45	0.35	6.83	9.04	2.89	1.52
10.195	0.032	0.032	235.30	0.00	11.21	10.03	4.91	1.55
11.185	b.d.		281.05	4.65	6.52		4.05	1.07

Appendix

Table A.2: Concentrations of BTEX, toluene, stable isotope values of toluene ($^{13}C/^{12}C$), PAHs, sulfide, sulfate, stable isotope values of sulfate ($^{34}S/^{32}S$ and $^{18}O/^{16}O$) and bacterial biomass in groundwater sampled with the high-resolution multi-level well in May 2009 (2nd week). av. = average, s.d. = standard deviation, b.d. = below detection limit.

Depth [m bls]	BTEX [mg L^{-1}]		Toluene [mg L^{-1}]		$\delta^{13}C$ [‰]	PAHs [mg L^{-1}]		SO_4^{2-} [mg L^{-1}]		$\delta^{34}S$ [‰]	$\delta^{18}O$ [‰]
	av.	s.d.	av.	s.d.		av.	s.d.	av.	s.d.		
6.385	5.14	3.13	2.13	1.35	-21.21	6.33		138.69	11.16	33.99	13.49
6.410	5.06	0.69	2.74	0.52	-22.06	7.30	2.01	145.29		33.83	13.61
6.435	3.70	0.76	2.07	0.37	-22.19	8.68	3.44	131.90	9.13	34.62	
6.460	9.98	5.45	5.74	3.03	-22.51	8.20	0.60	137.70	10.61	32.36	13.15
6.510	13.22	1.14	8.10	1.16	-22.57	3.76		149.62	10.18	28.47	13.45
6.535	10.95	3.57	7.53	2.59	-22.66	9.09	0.97	142.45	9.31	28.86	
6.560	21.27		14.84		-22.65	8.11	0.33	153.63	11.04	26.97	12.64
6.585	9.58	4.89	6.69	3.41	-22.48	7.77	1.13	161.37	11.30	26.15	
6.610	11.59	4.62	8.27	3.59	-22.52	5.11	4.96	163.50	9.84	25.85	12.65
6.635	14.15	10.15	10.16	7.21	-22.35	6.60	0.51	167.36	10.94	25.46	
6.665	14.74	0.64	10.56	0.72	-22.84	6.63	0.48	167.63	10.78	25.26	
6.695	6.73	3.77	4.92	2.95	-22.82	7.42	1.56	159.23	9.28	25.43	12.82
6.745	0.77	1.10	1.05		-23.35	5.82	0.07	176.62	9.48	24.86	12.75
6.805	0.83	1.18	1.17		-23.83	5.07	1.13	180.20	6.94	24.69	
6.855	0.53	0.75	0.77		-23.38	1.42	0.09	194.38	9.69	23.29	11.75
6.905	0.13	0.18	0.19		-23.14	0.88	0.17	203.82	13.96	21.35	
6.955	0.06	0.09	0.09		-23.55	1.06	0.05	217.42	14.77	19.64	11.35
7.005						1.43	0.06	212.41		18.42	
7.055						1.42	0.13	222.26		16.85	10.71
7.105						1.09		220.12		14.30	
7.205						0.58	0.11	240.54		11.18	9.63
7.255						1.16	0.04	257.49		9.77	
7.305						1.57	0.02	266.19		8.90	9.06
7.405						1.39	0.05	265.80		8.44	
7.505						0.60	0.03			8.79	
7.605						0.24	0.04	271.04	2.87	8.29	9.16
8.045						0.18	0.00	272.99	2.03	7.64	8.51
8.345						0.17	0.07	254.06	5.64	9.58	
9.045	0.15	0.21	0.27			0.20	0.06	279.67	4.32	4.35	8.30
10.195	0.00	0.00				0.40	0.01	239.98	6.26	9.49	9.15
11.185	0.08	0.02	0.07	0.02		0.14	0.03	280.19	1.87	4.52	

Table A.2 continued

Depth [m bls]	H_2S [mg L^{-1}] av.	s.d.	Cells [x 10^6 mL^{-1}] av.	s.d.
6.385	2.002	0.095	3.29	0.36
6.410	0.667	0.032	2.39	0.08
6.435	1.525	0.064	1.58	0.05
6.460	1.875	0.159	1.47	0.07
6.510	0.890	0.127	2.59	0.26
6.535	5.180	0.159	0.58	0.01
6.560	5.975	0.445	0.90	0.04
6.585	3.305	0.381	1.17	0.07
6.610	4.672	0.095	0.75	0.05
6.635	2.606	0.318	0.90	0.03
6.665	3.750	0.699	1.55	0.06
6.695	5.307	0.350	1.20	0.12
6.745	5.434	0.286	0.85	0.03
6.805	6.960	0.095	1.05	0.06
6.855	3.686	0.064	1.34	0.10
6.905	4.608	0.286	1.41	0.04
6.955	0.604	0.032	0.27	0.01
7.005	0.667	0.095	0.28	0.01
7.055	0.826	0.636	0.96	0.07
7.105	1.112	0.413	0.81	0.05
7.205	1.081	0.000	3.47	0.12
7.255	0.985	0.095	3.01	0.24
7.305	1.478	0.079	2.06	0.13
7.405	1.208	0.064	2.77	0.22
7.505	3.400	0.032	4.99	0.32
7.605	1.780	0.191	4.63	0.14
8.045	2.511	0.032	1.15	0.03
8.345	0.636	0.127	1.38	0.11
9.045	0	0	1.18	0.06
10.195	0	0	1.08	0.07
11.185	0.794	0.794	0.61	0.03

Table A.3: Concentrations of BTEX, toluene, PAHs, redox, sulfide, sulfate and bacterial biomass in groundwater sampled with the high-resolution multi-level well in 3rd week of May 2009. av. = average, s.d. = standard deviation, b.d. = below detection limit.

Depth [m bls]	BTEX [mg L^{-1}]		Toluene [mg L^{-1}]		PAHs [mg L^{-1}]		Redox [mV]	H$_2$S [mg L^{-1}]		SO$_4^{2-}$ [mg L^{-1}]		Cells [x 10^6 mL^{-1}]	
	av.	s.d.	av.	s.d.	av.	s.d.		av.	s.d.	av.	s.d.	av.	s.d.
6.385	4.92	0.82	1.92	0.41	3.03	0.32	-115	1.907	1.017	105.97	1.97	1.98	0.25
6.410	7.73	2.14	4.18	1.37	5.22	0.00	-118	1.621	0.095	111.38	0.50	2.29	0.14
6.435	8.74	1.42	4.91	0.80	5.09	1.01	-112	1.811	0.159	111.18	0.81	0.95	0.24
6.460	7.60	1.40	4.07	0.76	3.76	0.09	-87			123.89	0.15	0.82	0.09
6.510	11.13	1.87	6.34	0.76	3.05	0.34	-121	0.413	0.159	136.68	0.44	1.54	0.13
6.535	14.80	2.23	9.56	1.62	5.29	0.57	-137	1.049	0.159	129.36	1.32	0.80	0.08
6.560	16.77	2.58	11.86	1.77	2.32	0.72	-141	6.674	0.699	142.80	1.97	1.30	0.22
6.585	15.75	1.15	11.09	0.73	5.30	0.18	-121	3.877	0.254	147.90	2.38	1.10	0.32
6.610	19.46	3.34	14.26	2.30	3.80	0.87	-121	2.956	0.032	146.33	1.41	0.49	0.04
6.635	20.82	4.65	15.12	3.15	2.26	2.72	-111	3.845	0.222	152.48	5.72	0.72	0.06
6.665	19.60	0.69	15.42	0.47	5.25	0.10	-125	2.256	0.095	147.59	0.63	0.78	0.03
6.695	17.19	1.75	13.41	1.28	4.95	0.00	-126	3.559	0.127	137.00	0.79	2.77	0.23
6.745	15.72	0.37	12.74	0.19	4.63	0.33	-135	3.051	0.127	157.93	0.13	0.47	0.03
6.805	6.71	0.06	5.38	0.00	2.04	0.08	-154	3.718	0.095	164.25	2.27	0.47	0.02
6.855	2.47	0.27	1.98	0.19	1.78	0.31	-164	5.434	0.477	171.78	2.07	0.82	0.04
6.905	0.73	0.09	0.58	0.07	0.73	0.07	-155	4.703	0.191	181.43	10.24	0.53	0.05
6.955	0.34	0.02	0.27	0.02	0.61	0.10	-126	5.561	0.032	221.70	5.91	1.12	0.04
7.005	0.17	0.03	0.14	0.02	0.76	0.07	-110	0.222	0.032	239.06	0.07	0.39	0.04
7.055	0.09	0.02	0.06	0.01	0.38	0.05	-63			243.98	0.18	1.36	0.10
7.105	0.12	0.02	0.09	0.01	0.45	0.15	-51	0.032	0.032	231.13	8.31	0.53	0.04
7.205	0.11	0.07	0.09	0.06	0.24	0.05	-66			242.23	3.95	3.66	0.44
7.255	0.06	0.00	0.04	0.00	0.37	0.06		0.286	0.032	254.40	5.66	2.89	0.17
7.305	0.07	0.01	0.05	0.00	0.79	0.13		0.064	0.064	262.23	5.35	2.41	0.02
7.405	0.09	0.00	0.06	0.00	0.72	0.11		0.445	0.191	274.29	1.07	1.74	0.17
7.505	0.09	0.04	0.05	0.02	0.30	0.09	-105	0.191	0.191	265.97	1.68	3.41	0.18
7.605	0.09	0.02	0.03	0.01	0.08	0.09	-88	2.638	0.095	275.29	2.68	3.99	0.24
8.045	0.11	0.02	0.02	0.00	0.14	0.01	-81	0.572	0.254	277.14	89.11	2.54	0.14
8.345	0.23	0.02	0.02	0.01			-51	1.843	0.000	267.80	68.12	3.31	0.19
9.045	0.01	0.00	0.00	0.00			-53	0		282.40	5.14	1.32	0.03
10.195	0.02	0.01	0.02	0.01	0.63	0.27	-58	0		241.31	0.56	1.02	0.04
11.185	0.01	0.00	0.01		0.15	0.01	-59	0		280.87	2.31	0.55	0.04

Table A.4: Concentrations of BTEX, toluene, stable isotope values of toluene ($^{12}C/^{13}C$), PAHs, redox, sulfide, sulfate, stable isotope values of sulfate ($^{32}S/^{34}S$ and $^{16}O/^{18}O$) and bacterial biomass in groundwater sampled with the high-resolution multi-level well in June 2009 (1st week). av. = average, s.d. = standard deviation, b.d. = below detection limit.

Depth [m bls]	BTEX [mg L^{-1}] av.	s.d.	Toluene [mg L^{-1}] av.	s.d.	$\delta^{13}C$ [‰]	PAHs [mg L^{-1}] av.	s.d.	Redox [mV]
6.385	5.94	0.06	2.47	0.01	-20.51	8.65	3.97	33
6.410	7.23	0.19	3.63	0.16	-21.69	8.52	0.53	-11
6.435	8.14	1.44	4.01	0.29	-21.97	6.07		-14
6.460	8.78	1.30	5.22	0.69	-21.96			-21
6.510	15.29	0.19	7.80	0.13	-22.14			-133
6.535	16.26	0.77	10.33	0.45	-21.50	9.42	0.00	-126
6.585	18.25		14.87	0.77	-22.13	7.85	1.72	-11
6.610	20.30	1.15	14.27	0.34	b.d.			-48
6.635	20.12	7.22	18.33	4.54	-22.21	10.60		37
6.665	19.62	3.02	13.25	1.78	-22.32	9.04	0.40	-98
6.805	10.19	0.35	5.97	0.02	b.d.	5.51		47
6.855	2.05		1.87	0.01	-22.84	4.12	0.87	-127
6.905	0.74		0.65	0.04	b.d.	1.86	0.38	-107
6.955	0.71	0.46	0.33	0.02	-22.21	1.83	0.25	-13
7.005	0.41	0.08	0.16	0.06	b.d.	1.91	0.00	-40
7.055	0.15	0.01	0.09	0.00	-22.13	1.91		49
7.105	0.18	0.01	0.13	0.00		0.97		34
7.205	0.12	0.02	0.08	0.00		0.50	0.03	51
7.255	0.11	0.02	0.07	0.01		0.89		-45
7.305	0.12	0.00	0.09	0.00		2.05	0.06	-37
7.405	0.11	0.01	0.08	0.01		1.94	0.14	-2
7.455	0.11	0.02	0.07	0.01				
7.505	0.12	0.01	0.08	0.01		0.26	0.00	-23
7.605	0.11	0.01	0.06	0.00		0.01	0.01	-46
8.045	0.17	0.02	b.d.			0.10	0.01	-43
8.345	0.13	0.05	b.d.					-64
9.045	0.06	0.01	0.05	0.00		0.00	0.00	-57
10.195	0.01	0.00	b.d.			0.20	0.05	28
11.185	b.d.		b.d.			0.07	0.05	-44

Table A.4 continued

Depth [m bls]	H$_2$S [mg L^{-1}] av.	s.d.	SO$_4^{2-}$ [mg L^{-1}] av.	s.d.	δ^{34}S [‰]	δ^{18}O [‰]	Cells [x 10^6 mL^{-1}] av.	s.d.
6.385	6.896	0.095	102.06	3.40	39.79	14.76	2.73	0.15
6.410	2.097	0.000			37.94	14.79	2.77	0.23
6.435	3.591	0.032	118.71	1.05	36.62	14.53		
6.460	3.210	0.032	130.86	1.78	33.23	14.07	3.13	0.26
6.510	3.273	0.095	143.67	1.59	30.45	13.70	2.47	0.16
6.535	6.769	0.032	137.34	1.59	31.08	13.77	1.93	0.11
6.585	2.479	0.191			27.97	13.31		
6.610	1.970	0.191	152.28	3.37	28.00	13.57	0.85	0.07
6.635	2.129	0.222	153.28	1.14	27.33	13.03	1.19	0.17
6.665	1.653	1.017	156.38	0.89	26.16	13.05	1.07	0.13
6.805	3.972	0.222	168.19	0.95			0.99	0.12
6.855	6.165	0.064	173.12	0.32			1.29	0.05
6.905	6.070	0.095	194.48	1.87	25.16	14.27	1.71	0.05
6.955	5.498	0.032	223.40	3.05	22.35	12.63	1.57	0.15
7.005	0.794	0.032	241.24	2.99	18.20	12.67	1.29	0.03
7.055	0.604	0.095	243.82	2.73	15.60	11.23	0.95	0.04
7.105	b.d.		222.07	2.53	14.60	12.15	1.64	0.07
7.205	b.d.		232.76	0.53	14.76	11.94	3.44	0.04
7.255	0.222	0.159	245.47	2.31	12.67	10.68	3.78	0.33
7.305	b.d.		256.46	1.43	10.84		2.62	0.16
7.405	0.318	0.064	271.85	3.73	9.44	10.51	2.04	0.08
7.455	0.222	0.222	272.01	1.69	7.84	9.39		
7.505	0.540	0.032	271.12	4.21			4.23	0.03
7.605	3.210	1.239	284.24		8.33	9.32	2.43	0.10
8.045	0.540	0.032	287.53	1.27	7.44	10.80	0.77	0.04
8.345	1.653	0.064	267.49	1.46	7.47	10.33	0.60	0.02
9.045	b.d.		282.82	2.30	9.19		0.44	0.03
10.195	0.095	0.350	250.41	0.15	5.07		0.97	0.01
11.185	b.d.		288.10	1.32	5.03	9.58	0.47	0.03

Table A.5: Concentrations of the detected EPA-PAHs Naphthalene, Acenaphthene and Fluorene in groundwater sampled with the high-resolution multi-level well in September 2008, b.d. = below detection limit.

Depth [m bls]	September 2008 Naphthalene [mg L^{-1}]	Acenaphthene [mg L^{-1}]	Fluorene [mg L^{-1}]
6.385	3.28	0.17	0.11
6.410	3.97	0.23	0.24
6.435	4.18	0.21	0.02
6.460	4.74	0.22	0.05
6.510	5.13	0.29	0.19
6.535	5.07	0.24	0.14
6.585	4.99	0.31	0.09
6.610	4.84	0.21	0.11
6.635	5.40	0.26	0.14
6.665	5.22	0.33	0.17
6.695	5.32	0.30	0.14
6.745	1.89	0.35	0.19
6.805	0.92	0.44	0.23
6.855	1.65	0.39	0.22
6.905	1.21	0.60	0.30
6.955	0.73	0.84	0.32
7.005	0.24	0.78	0.44
7.055	0.06	1.20	0.67
7.105	0.02	1.24	0.64
7.155	0.02	1.31	0.64
7.205	0.01	1.63	0.79
7.255	0.01	1.58	0.79
7.405	0.01	1.74	0.75
7.505	0.03	1.64	0.66
7.605	0.02	1.57	0.64
8.045	b.d.	0.51	0.30
9.045	b.d.	0.42	0.36
10.195	b.d.	0.78	0.34
11.185	b.d.	0.46	0.10

Table A.6: Concentrations of the detected EPA-PAHs Naphthalene, Acenaphthene and Fluorene in groundwater sampled with the high-resolution multi-level well in May 2009 and June 2009, b.d. = below detection limit.

May 2009				June 2009			
Depth [m bls]	Naphthalene [mg L^{-1}]	Acenaphthene [mg L^{-1}]	Fluorene [mg L^{-1}]	Depth [m bls]	Naphthalene [mg L^{-1}]	Acenaphthene [mg L^{-1}]	Fluorene [mg L^{-1}]
6.385	5.94	0.49	0.18	6.385	5.16	0.43	b.d.
6.410	7.23	0.46	0.12	6.410	5.00	0.40	b.d.
6.435	8.14	0.31	b.d.	6.435	4.59	0.49	b.d.
6.460	8.78			6.460	5.92	0.33	b.d.
6.510	15.29			6.510	6.07	0.23	b.d.
6.535	16.26	0.29	b.d.	6.535	5.70	0.95	b.d.
6.585	18.25	0.15	0.04	6.560	7.54	0.32	
6.610	20.30			6.585	6.81	0.21	b.d.
6.635	20.12	b.d.	b.d.	6.610	6.72	0.13	b.d.
6.665	19.62	0.33	0.06	6.635	6.74	0.28	0.01
6.805	10.19	0.51	0.31	6.665	7.23	0.31	0.02
6.855	2.05	1.26	0.65	6.695	4.94	0.44	b.d.
6.905	0.74	1.14	0.49	6.745	5.23	0.45	0.02
6.955	0.71	1.18	0.62	6.805	3.18	1.02	0.05
7.005	0.41	1.25	0.65	6.855	1.04	0.77	0.09
7.055	0.15	1.43	0.48	6.905	0.22	0.74	0.10
7.105	0.18	0.81	0.16	6.955	0.10	0.89	0.17
7.205	0.12	0.50	b.d.	7.005	b.d.	1.16	0.27
7.255	0.11	0.70	0.18	7.055	b.d.	1.14	0.29
7.305	0.12	1.35	0.70	7.105	b.d.	0.87	0.22
7.405	0.11	1.30	0.64	7.205	b.d.	0.56	0.02
7.455	0.11			7.255	b.d.	0.94	0.22
7.505	0.12	0.26		7.305	b.d.	1.24	0.33
7.605	0.11	0.01		7.405	b.d.	1.11	0.28
8.045	0.17	0.10		7.505	b.d.	0.51	0.09
8.345	0.13	0.07		7.605	b.d.	0.24	b.d.
9.045	0.06	0.00		8.045	b.d.	0.18	b.d.
10.195	0.01	0.20		9.045	b.d.	0.20	b.d.
11.185		0.07		10.195	b.d.	0.40	b.d.

Table A.7: Localization and BTEX concentration of conventional multi-level wells located at the Düsseldorf site sampled and analyzed by the Stadtwerke Düsseldorf.

Well	19209	19185	19214	19201	19202
x-value	57122.15	57181.02	57211.09	57202.48	57122.15
y-value	76719.71	76703.61	76719.62	76699.97	76707.08
Depth [m bls]	May 2006 BTEX [µg L^{-1}]				
6			240		113
6.5			429	34776	683
7	186	138	16	2014	19
8	2594	638	8	484	6
9	90	64	1	520	2
10	15	29	15	31802	2
	Feb 2007 BTEX [µg L^{-1}]				
6.5			475	6317	
7	0		163	4823	26
8	0	56	15	187	0
9	0	1632	0	143	0
10	0	7	0	79	0
	Aug 2008 BTEX [µg L^{-1}]				
6.5			0	7695	324.3
7			0	218	21.5
8	539.2	523	0	186.8	18.9
9	89.16	124.82	4.4	50.6	6.31
10	17	38.7	0	41.1	12.5
	Aug 2009 BTEX [µg L^{-1}]				
7	17.62	69.05		279	
8	168.94	429		256.6	
9	194.66	155.42		12.38	
10	17.74	6.41		11.9	

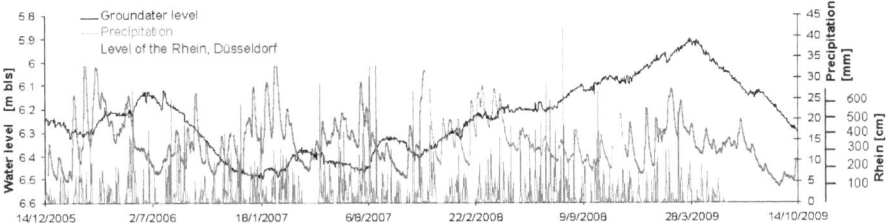

Figure A.1: Groundwater level at a neighboured well of the high-resolution multi-level well as well as precipitation (www.dwd.de) and level of the river Rhein in Düsseldorf.

Determination of initial toluene concentrations

The modeled source value of $\delta^{13}C$ of toluene at the investigated aquifer is -25‰ (Prommer at al. 2009), thus I used it for the calculation of the amount of biodegradation and the initial toluene concentration. The isotope fractionation factor α = 0.9983 for toluene degradation for various strains (Meckenstock et al. 1999) was used. R_0 is the $\delta^{13}C$ value of the non-degraded toluene and R_t the measured $\delta^{13}C$, both multiplied with 1000 (Rayleigh 1896, Meckenstock et al 2004).

$$\text{Calculated amount of biodegradation [\%]} = 1 - \left[\left(\frac{R_t}{R_0}\right)^{\frac{1}{\alpha-1}}\right] \times 100$$

With the measured toluene, the initial concentration of toluene was calculated. An additional method was used for determination of the initial toluene concentration. It was assumed that the difference of the background sulfate and the measured sulfate was completely used for toluene biodegradation (Anneser et al. 2008). For the degradation of one mol toluene 4.5 moles sulfate are needed.

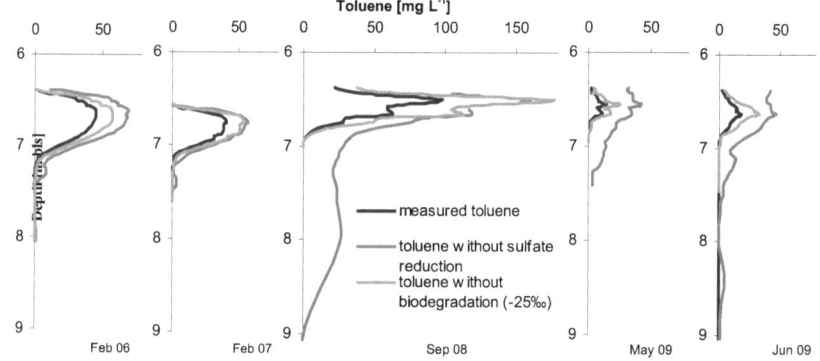

Figure A.2: Vertical distribution of measured toluene, initial calculated toluene neglecting sulfate reduction and initial calculated toluene with stable isotope neglecting biodegradation.

References:

Anneser B, Einsiedl F, Meckenstock RU, Richters L, Wisotzky F, Griebler C (2008) High-resolution monitoring of biochemical gradients in a tar oil-contaminated aquifer. Applied Geochemistry 23:1715-1730

Meckenstock RU, Morasch B, Warthmann R, Schink B, Annweiler E, Michaelis W, Richnow HH (1999) $^{13}C/^{12}C$ isotope fractionation of aromatic hydrocarbons during microbial degradation. Environmental microbiology 1:409-414

Meckenstock RU, Morasch B, Griebler C, Richnow HH (2004) Stable isotope fractionation analysisi as a tool to monitor biodegradation in contaminated aquifers. Journal of Contaminant Hydrology 75:215-255

Prommer H, Anneser B, Rolle M, Einsiedl F, Griebler C (2009) Biochemical and Isotopic Gradients in a BTEX/PAH Contaminant Plume: Model-Based Interpretation of a High-Resolution Field Data Set. Environmental Science & Technology 43:8206-8212

Rayleigh L (1896) Theoretical considerations respecting the separation of gases by diffusion and similar processes. Philosophical Magazine Series 5 42:259, 493-498

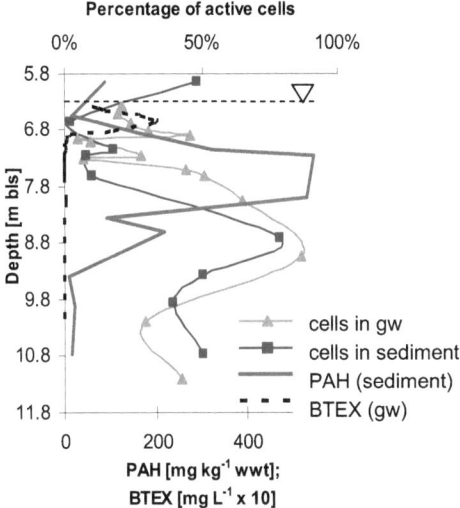

Figure A.3: Vertical distribution of the percentage of active (via amount of ATP) to total cells in water (gw) and sediment sampled in June 2009.

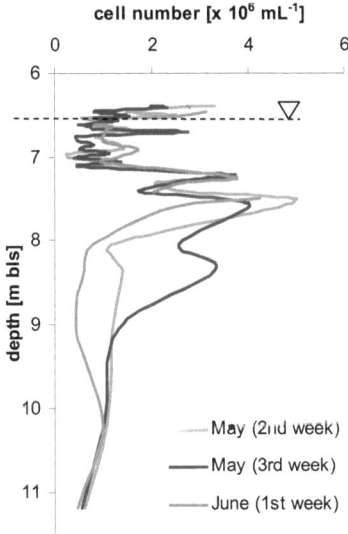

Figure A. 4: Short-term comparison of the vertical distribution of the total cell number in groundwater.

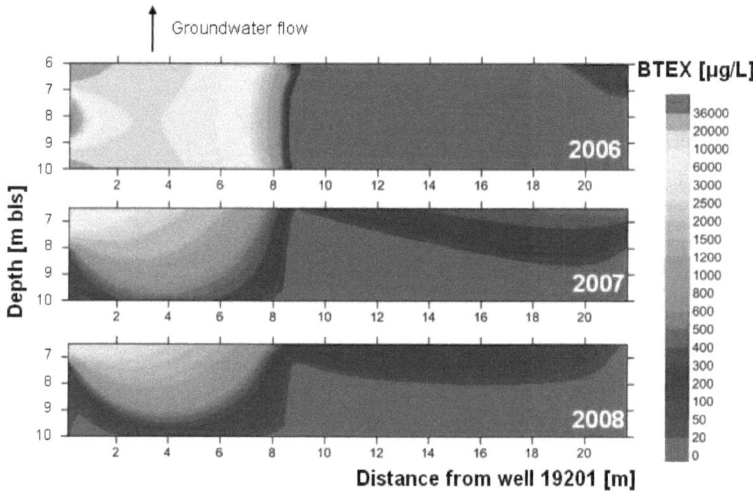

Figure A. 5: Calculated width of the contaminant plume with Surfer 9 software and BTEX concentrations of wells sampled and analyzed b the Stadtwerke Düsseldorf. Unfortunately, not enough data were available for the calculation of the plume width in 2009.

Authorship clarifications

CHAPTER 2: **"High resolution analysis of contaminated aquifer sediments and groundwater – what can be learned in terms of natural attenuation?"**
The PhD candidate performed the analysis of the total bacterial cell numbers, wrote the parts in the manuscript belonging to bacterial cell numbers and created figure 2.3. The idea and concept for the experimental design were carried out by Dr. Bettina Ruth (nee Anneser) and Dr. Christian Griebler. Dr. Giovanni Pilloni performed the microbial fingerprinting, calculated the statistics and evaluated the results together with Dr. Tillmann Lueders. Dr. Bettina Ruth, Dr. Florian Einsiedl and Dr. Christian Griebler performed the remaining analyses and evaluated the results related to the on site geochemistry. The manuscript was published on *Geomicrobiology Journal*, 27:130-142 (2010).

CHAPTER 3: **"Collapse and recovery of intrinsic toluene degradation – transient hydraulic conditions control natural attenuation in a tar oil contaminated sandy aquifer"**
The idea and concept for the experimental design were carried out by the PhD candidate and Dr. Christian Griebler. The sampling in 2008 as well as all short-term samplings in 2009 were organized and performed by the PhD candidate. All analyses of these samples were done by the PhD candidate. Dr. Bettina Ruth (nee Anneser) performed the samples of 2005 to 2007. The manuscript was written by the PhD candidate, the results were evaluated together with Dr. Bettina Ruth and Dr. Christian Griebler, and the manuscript was prepared for submission to *Environmental Science & Technology*.

CHAPTER 4: **"Quantification and preservation of aquifer sediment bacteria – a multiple assay comparison"**
First bacterial cell counts were done by Regina Drexel. She compared the different detachment procedures, dyes, fixation protocols, used density gradient centrifugation and compared microscopy with flow cytometry. The PhD candidate compared the influence of different storage, performed the sampling and the vertical depth profile of the natural sediment, determined the correction factor, ATP content, improved the flow cytometric measurements and performed the statistics. Katrin Hörmann performed the DNA extraction and PCR. The manuscript was written by the PhD candidate and submitted by the corresponding author Dr. Chrisitian Griebler to *Aquatic Microbial Ecology* on 29.06.2011.

Publications

B. Anneser, G. Pilloni, **A. Bayer**, T. Lueders, C. Griebler, F. Einsiedl and L. Richters, **High Resolution Analysis of Contaminated Aquifer Sediments and Groundwater – What Can be Learned in Terms of Natural Attenuation?**, *Geomicrobiology Journal*, 27 (2): 130-142 (2010).

A. Bayer, B. Ruth and C. Griebler, **Collapse and recovery of intrinsic toluene degradation – transient hydraulic conditions control natural attenuation in a tar oil contaminated sandy aquifer**, *Environmental Science & Technology* (in preparation).

A. Bayer, R. Drexel, K. Hörmann, T. Lueders and C. Griebler, **Quantification and preservation of aquifer sediment bacteria – a multiple-assay comparison**, *Aquatic Microbial Ecology* (submitted on 26.06.2011).

G. Pilloni, **A. Bayer**, B. Ruth, M. Engel, C. Griebler and T. Lueders, **Ecological feedbacks between hydraulic dynamics and natural attenuation in a contaminated aquifer**, *Nature Goescience* (submitted on 07.09.2011).

Acknowledgement

Already during my diploma thesis wondering about what is coming next, I thought about doing my PhD in the institute of groundwater ecology. When a former colleague, and now a good friend, Kristina Pfannes applied for a postdoc position in Rainer Meckenstock's group, I told her she will get this job and then she can help me to get a PhD in this institute. I was not sure at all that it can work, but first Kristina was lucky and got the position, then I was lucky.

Christian Griebler made it possible to do my PhD in his group and gave me this fruitful project. I want to thank him for allowing me working on the one hand independently but on the other hand he had always an open ear and many helping ideas when I had questions or problems. I am very glad that I got in this research field and in his working group doing a very good combination of samplings in the field and analysis in the lab.

Furthermore, I want to recognize Prof. Dr. Rainer Meckenstock for the chance to do this work at the IGÖ and for his advice and support. I am also grateful to Prof. Dr. Arnulf Melzer who agreed to be second supervisor within a very short time.

Furthermore, I want to thank Bettina Ruth that I could work further on this project and use a lot of things and methods she tried out. I always could ask her, she helped with the sampling and reading parts of my manuscript.

Thanks a lot to Giovanni Pilloni for the fruitful and very good teamwork with the Düsseldorf-project and samplings in Düsseldorf. My thesis will not exist without the sampling campaigns in Düsseldorf and the many helping hands of Giovanni, Betti, Katrin, Agnieszka, Tina, Christian, Tillmann and especially of Lars Richters who made the access to the Düsseldorf site possible and provided me a lot of data evaluated by the Stadtwerke Düsseldorf.

Special thanks to Tina Höche for introducing me and helping a lot with measuring stable isotopes as well as Martin Elsner for using the GC-C-IRMS-machine and helping with analyzing the data. Furthermore, Harald Lowag measured the stable isotope data of the sulfate and handed over always perfect results. Michael Stöckl and Günter Teichmann helped me anytime with some measurements and when my computer had a problem.

The nice time I spent in the office would not be possible with my room mates Marko, Giovanni, Katrin and Clemens. Thank you for the discussions about science as well non-science things, laughing, arguing, planning trips to Biergarten,…and co-bearing the changes of the diploma, bachelor, master and "what-ever" students in our office as well as the noise of the many reconstructions.

For reading and correcting my English writing thanks go to Susanne, and also for her encouragement.

Without my parents, who facilitates my education and were always be there for me this work would not be possible. Thanks go also to my sisters, being the best younger sister ever and giving me the feeling that there is always a way!

Finally, my warmest thanks belong to Christoph who accompanied me through bad and good times during my PhD. Thank you for everything.

i want morebooks!

Buy your books fast and straightforward online - at one of world's fastest growing online book stores! Environmentally sound due to Print-on-Demand technologies.

Buy your books online at
www.get-morebooks.com

Kaufen Sie Ihre Bücher schnell und unkompliziert online – auf einer der am schnellsten wachsenden Buchhandelsplattformen weltweit! Dank Print-On-Demand umwelt- und ressourcenschonend produziert.

Bücher schneller online kaufen
www.morebooks.de

VDM Verlagsservicegesellschaft mbH
Heinrich-Böcking-Str. 6-8   Telefon: +49 681 3720 174   info@vdm-vsg.de
D - 66121 Saarbrücken   Telefax: +49 681 3720 1749 www.vdm-vsg.de

Printed by Books on Demand GmbH, Norderstedt / Germany